Agroforestry For Sustained Productivity in Arid Regions

Agroforestry For Sustained Productivity in Arid Regions

Editors

J.P. Gupta | B.M. Sharma
Central Arid Zone Research Institute
Jodhpur - 342 003, INDIA

SCIENTIFIC PUBLISHERS

Published by:
Scientific Publishers
5A, New Pali Road, P.O. Box 91
Jodhpur 342 001, India
E-mail: info@scientificpub.com
Website: www.scientificpub.com

Print : 2026

ISBN: 978-81-72331-55-9 (Hardbound)
ISBN: 978-93-87991-03-3 (E-book)

Printed in India

FOREWORD

Arid lands occupy about 12% geographical area of the country. Inspite of harsh environment and low crop production, these are inhabited by vast population of human beings and livestock. The livestock population (ACUS) is expected to be 11.3 million (about 17.6% over 1983 level) and the fodder requirement will be around 28.2 m tonnes in 2000 AD. To meet this increasing demand there is over-exploitation leading thereby to desertification and degradation of land, water and vegetation resources. In view of this there is an urgent need for developing and use of integrated farming systems for efficient utilization of land and water resources and sustained production. Agroforestry, silvi-pasture, agri-horticulture are some such systems which can provide stability and sustainability to the region.

Dr. J.P. Gupta and Dr. B.M. Sharma have compiled information on these aspects in the form of a book entitled, "*Agroforestry for sustained productivity in arid regions*". I am sure this will be of great use to the scientists and other agencies engaged in development of the arid areas.

(R.S. Paroda)
Secretary, DARE
Government of India
and
Director General
Indian Council of Agricultural
Research New Delhi

PREFACE

Agroforestry is an integrated land use system combining trees, crops and/or animals for sustained production of grains, fuel wood, fodder and other products of economic value. This system had been *in vogue* in arid areas since time immemorial possibly for efficient utilization of meagre and off season rainfall, nutrients and also for providing economic security and environmental protection. In recent years, arid regions registered an alarming increase in population of human beings and livestock. This has put tremendous pressure on agricultural lands, forest lands and grazing lands. As a result of this even the marginal and submarginal lands are put under cultivation. There is deforestation and degradation of grazing lands for meeting fuel wood and fodder needs. The net result is degradation and desertification. These processes need to be halted and reversed. For achieving this objective, Central Arid Zone Research Institute (CAZRI), Jodhpur initiated work on the development of different agroforestry systems for different land types and agro-ecological sub-regions. Lot of valuable information has been obtained on various aspects of agroforestry systems, improved genotypes of trees and grasses, soil and water conservation, nutrient-dynamics, socio-economic conditions and transfer of technology. This information has been compiled in the form of this publication. It is hoped that the information contained in this volume will be valuable to researchers, extension workers and all others engaged in development of the arid regions.

J.P. Gupta
Head of Division
&
B.M. Sharma
Sr. Scientist,
Division of Resource Management
Central Arid Zone Research
Institute, Jodhpur-342 003

THE CONTRIBUTORS

R.K. Aggarwal
CAZRI, Jodhpur

R. Raj. Bhansali
CAZRI, Jodhpur

L.P. Bharara
CAZRI, Jodhpur

T.K. Bhati
CAZRI, Jodhpur

H.C. Bohra
CAZRI, Jodhpur

R.K. Goyal
CAZRI, Jodhpur

J.P. Gupta
CAZRI, Jodhpur

B.L. Jain
NBSS & LUP, Udaipur

S.K. Jindal
CAZRI, Jodhpur

D.C. Joshi
CAZRI, Jodhpur

S. Kathju
CAZRI, Jodhpur

Praveen Kumar
CAZRI, Jodhpur

B.K. Mathur
CAZRI, Jodhpur

P.R. Ojasvi
CAZRI, Jodhpur

M.P. Rajora
CAZRI, Jodhpur

A.S. Rao
CAZRI, Jodhpur

A.V. Rao
CAZRI, Jodhpur

S.S. Rathore
CAZRI, Jodhpur

S.K. Saxena
CAZRI, Jodhpur

A.K. Sharma
CAZRI, Jodhpur

B.M. Sharma
CAZRI, Jodhpur

S.K. Sharma
CAZRI, Jodhpur

K.C. Singh
CAZRI, Jodhpur

M.P. Singh
CAZRI, Jodhpur

R.N. Singh
CAZRI, Jodhpur

K.R. Solanki
NRCAF, Jhansi

J.C. Tewari
CAZRI, Jodhpur

M.S. Yadav
CAZRI, Jodhpur

CONTENTS

	Foreword	(v)
	Preface	(vii)
	List of Contributors	(ix)
1.	Arid zone soils and their suitability for agroforestry -D. C. Joshi	1
2.	Climatic features and their modifications through agroforestry - A. S. Rao	13
3.	Traditional agroforestry systems in western Rajasthan - S. K. Saxena	21
4.	Some alternative production systems and their management for sustainability -- J. P. Gupta	31
5.	Some improved silvipastoral systems for sustainable production -- S. K. Sharma	41
6.	Integrated farming systems for higher and sustained productivity in Watersheds of arid areas -TK. Bhati	47
7.	Shelterbelt Management for environmental improvement in hot arid ecosystem -A. K. Sharma and J. P. Gupta	55
8.	Management of degraded grasslands in arid region of Rajasthan - K. C. Singh	61
9.	Nutrient cycling in agroforestry - B. M .Sharma	69
10.	A conceptual model for the evaluation of agroforestry systems -P. R. Ojasvi and R. K. Goyal	79
11.	Role of legumes alongwith biofertilizers in agroforestry systems for higher plant productivity -A .V. Rao	89

12. Water relations of desert vegetation
- S. Kathju 99

13. Soil fertility management in agroforestry systems
- R. K. Aggarwal and Praveen Kumar 107

14. Feed resources of Indian arid zone 2 Their nutritive value for the livestock
- H. C. Bohra and B. K. Mathur 113

15. Potential role of agroforestry in arid zone in 2000 AD
- B. M. Sharma and J. P. Gupta 117

16. Insect pest management in agroforestry
-M. P. Singh 125

17. Diseases of desert trees and their management
- R. Raj Bhansali and S. K. Jindal 133

18. Important fodder trees of arid zone and their management .
- J. C. Tiwari 147

19. Genetic improvement of tree species suitable for arid zone
-KR. Solanki 155

20. Improved varieties of pasture grasses and their cultivation
- M. S. Yadau and M. P Rajora 165

21. Social issues in agroforestry land use systems on arid lands
-L. R Bharara 169

22. Development of degraded pasturelands in arid region- A case study
- S. S. Rathore B. L. Jain and B. M. Sharma 183

23. Transfer of agroforestry technology in arid zone- A case study
- R. N. Singh 193

1

Arid Zone Soils and their Suitability for Agroforestry

D.C. Joshi

Soil fulfils three essential requirements for plant growth viz. supply of moisture, supply of nutrients and mechanical support to plants. Therefore, to establish an agroforestry system a sound foundation in terms of good soil is a pre-requisite. Besides common characteristic of aridity, the soils of arid region of Rajasthan vary widely in terms of texture, salinity, calcareousness, depth and underlying substrata which have direct influence on their biological productivity. These variations in soil characteristics are attributed to parent material, landscape attributes and the micro relief (Dhir, 1977). Based on their salient characteristics the major soil groups recognised in arid and semi-arid regions of Rajasthan are described below. Physico- chemical characteristics of the soil profiles are reported in Table 1.

SOILS OF ARID AND SEMI-ARID REGIONS OF RAJASTHAN

Dune and Interdunal Soils

Very deep sandy soils are associated with dune and inter-dunes, covering 31 per cent area of arid Rajasthan. Large stretches of sand dunes occur in western part of Barmer, Jaisalmer, Bikaner and Churu districts. These sand dunes are mostly parabolic in shape but longitudinal, transverse and barchan types of dunes also occur. Except barchans, all other types of dunes are 30 to 50 m high, mostly stable and have thin cover of the loose sand and scanty vegetation. The barchan dunes are crescent shaped, 5 to 7 m high, devoid of vegetation and covered with a mass of loose sand which is mobile.

TABLE 1. Physical and chemical characteristics of arid zone soils

Depth (cm)	Particle size distribution (%)				$CaCO_3$ (%)	Moisture equivalent (%)	pH	Electrical conductivity (dSm^{-1})
	Clay	Silt	Fine sand	Coarse sand				
			DUNE AND INTERDUNE SOILS					
0-25	1.9	1.1	96.0	4.5	1.5	2.0	8.04	0.16
25-50	4.8	1.5	86.9	4.3	1.4	3.5	8.01	0.16
50-75	4.2	1.4	89.9	3.5	1.5	3.0	8.04	0.16
75-100	5.5	1.5	88.8	3.5	1.3	4.5	8.10	0.16
			LIGHT BROWN SANDY SOILS					
0-20	3.8	3.8	88.6	10.0	0.8	4.8	9.80	0.24
20-40	9.6	16.3	64.3	50.0	4.9	15.8	7.45	0.66
40-60	13.0	14.1	58.2	76.2	8.3	17.1	7.54	0.26
			HARD PAN SOILS					
0-10	7.2	6.8	74.7	11.3	nil	7.8	8.21	0.10
10-30	9.3	8.3	69.3	12.6	0.2	10.3	8.24	0.10
30-70	10.4	9.2	67.1	12.0	1.3	10.5	8.20	0.10
Below 70	13.2	11.6	22.3	14.3	38.6		8.30	0.30

Contd.........

				BROWN LIGHT LOAM SOILS				
0-20	8.7	6.2	59.7	24.8	0.1	7.4	8.70	0.30
20-40	9.6	13.8	58.7	17.9	0.1	12.2	8.55	0.30
40-70	13.1	13.5	58.5	13.6	1.4	13.0	8.60	0.30
70-80	11.3	10.4	58.0	12.4	1.9	12.6	8.65	0.30
				GRAY BROWN LOAM SOILS				
0-20	22.6	9.8	54.4	12.4	0.30	10.9	8.40	0.30
20-60	25.9	14.9	42.5	15.9	0.49	14.4	8.40	0.20
60-80	25.3	11.3	49.4	12.2	2.04	10.9	8.40	0.20
80-100	26.4	10.4	47.5	10.4	5.30	11.5	8.50	0.20
				SALT AFFECTED SOILS				
0-20	28.7	29.1	30.3	0.5	11.3	30.1	7.07	28.6
20-50	16.4	24.9	49.0	0.6	8.5	27.5	7.23	19.9
50-70	16.6	24.9	51.0	0.8	6.7	28.6	7.44	21.3
70-100	24.6	28.3	40.0	0.4	6.8	36.1	7.30	26.1

The soils associated with dunes are uniformly sandy, non-coherent and pale brown to light yellowish brown (10 YR 6/3 to 6/4) up to 150 cm depth. The silt and clay content range from 0.4 to 1.3 per cent and 1.8 to 4.5 per cent, respectively. The $CaCO_3$ content (0.6 to 5.0 per cent) is present mostly in diffuse form. The available water capacity of these soils is very low, one metre deep soil retains 50 to 70 mm available water. Wind erosion and deep percolation losses are major problems associated with these soils. The interdune soils have slightly more silt, clay and organic matter content. The available water capacity and nutrient status of interdune soils are better than dune soils. These soils have been classified as sandy, mixed, hyperthermic, Torripsamments.

Light Brown Sandy Soils

These are the soils of sandy plain having scattered sand hummocks and occasional dunes on the surface. Soils of this group occur in all arid zone districts and cover almost 34 per cent area of western Rajasthan. Surface soils, are loamy sand/fine sand, brown to yellowish brown (10 YR 5/3, 5/4) and very weakly aggregated; having clay and silt content 3.6 to 6.2 per cent and 1.8 to 3.2 per cent, respectively. The subsoil is loamy sand, weak sub- angular blocky with slightly more silt and clay. Below 60 to 90 cm depth lime concretionary layer occurs which is impervious to roots and water. The available water capacity of one metre profile ranges from 60 to 80 mm. These soils are the members of coarse loamy, mixed, hyperthermic family of Camborthids/Calciorthids.

Hard Pan Soils

Soils of this group are shallow (30 to 40 cm deep) and very compact lime concretionary strata is present below this layer. These soils are pale brown to yellowish brown (10 YR 6/2 t' 5/4), fine sand to loamy sand with single grain to weakly granular structure. Clay and silt content vary from 6.1 to 8.4 per cent and 4 to 5.5 per cent, respectively. The lime concretionary layer has been compacted due to lime and silica and at places strata is exposed to the surface. It is impervious to roots and water. Such hard pan soils are found near Phalodi (Jodhpur), Jayal- Katoti (Nagaur) and in parts of Jaisalmer and Barmer districts. Because of shallow depth the moisture retention capacity of the profile is very low (30 to 50 mm). These soils dry quickly and do not support perennial vegetation particularly trees. According to soil taxonomy,the soils have been classified as coarse loamy, mixed, hyperthermic, Lithic Paleorthids.

Brown Light Loam Soils

These soils are dark brown to dark reddish brown (7.5 YR 4/4, 5 YR 5/4), 60 to 150 cm deep and developed from prolluvium. At surface, the soils are loamy fine sand which changes to sandy loam and loam in sub soil. The silt and clay contents in the sub soil range between 13.8 to 14.4 per cent and 7.3 to 8.3 per cent, respectively. Soils of this group occur mainly in Jodhpur, Nagaur, Barmer and Jalor districts. The soils are generally non calcareous and have gravelly layer thinly coated with lime at 90 to 120 cm depth. Available water capacity of these soils vary from 80 to 120 mm per metre depth. These soils are very potential, both under rainfed and irrigated conditions. According to soil taxonomy, the soils have been classified as coarse loamy, mixed hyperthermic, Camborthids.

Gray Brown Loam Soils

These soils have developed from the fine textured alluvium, deposited by a number of streams originating from the Aravalli hill range and flowing through plain area to join the river Luni. These soils occur in parts of Nagaur, Jodhpur, Pali and Jalor districts. The soil profile has gray brown to dark grayish brown (10 YR 5/2, 3/2) loam, clay loam and silt clay loam texture. The silt and clay content vary from 11 to 14 per cent and 15 to 23 per cent, respectively. The depth of soil vary from 60 to 120 cm below which a semi compact lime concretionary zone occurs. Available water capacity of the soil vary from 130 to 200 mm per metre depth. Moderate to severe water erosion is major problem on uplands. Natural salinity is also of common occurrence in this soil group. These soils belong to fine loamy, mixed, hyperthermic family of Camborthids/ Calciorthids/ Salorthids.

Salt Affected Soils

In the arid part of Rajasthan, salinity, both in soils and ground water, assumes unusual proportion and poses a problem to reckon with. Natural saline flats, saline depression, relict saline and secondary salinised due to high water table and saline/sodic due to irrigation with saline/sodic water are of common occurrence. Salt affected soils occur in scattered small patches in all the arid districts but their larger extent is met with in the south eastern part of Jodhpur, Nagaur, Pali and south western part of Jalor district. Other larger saline tract occurs in the flood plain of Ghaggar river in Hanumangarh, Sriganganagar and Bikaner districts.The soils are grayish brown to dark grayish brown (10 YR 5/2, 4/2), loam, clay loam and silt clay loam and 60 to 150 cm deep. The soil profile is charged with high salinity (EC_2 13 to 84 dSm^{-1}). Salinity in the surface soil is high and decreases with depth (Kolarkar *et al.*, 1980). Because of high salinity and sodicity, the soil profile turns very compact, soils develop prismatic/columnar structure. Because of adverse physical conditions, the infiltration and water permeability of profile are restricted. Inspite of high salinity, nutrient status of these soils is comparable with the associated normal soils (Joshi *et al.*, 1988a, 1988b).

The secondary saline/sodic soils due to irrigation with poor quality water are widespread in the irrigated areas. The saline water (EC 3 to 10 dSm^{-1}) irrigated soils develop high salinity (EC_e 8 to 10 dSm^{-1}). The salinity build up after one irrigated *rabi* crop is leached during subsequent rainy season. High RSC water (5 to 20 meq L^{-1}) irrigated soils develop sodicity (pH 8.9 to 9.5) and adverse physical conditions (Joshi and Dhir, 1994).

Shallow Rocky Gravelly Soils

In the vicinity of hills, uplands and sheet rock areas, the soils are shallow having thin layer of soil (15 to 20 cm) mixed with gravels or rock fragment. Soil erosion due to rain water is very active on these lands. At many places such areas are cultivated or kept under pastures. Because of shallow rooting zone, very low stored moisture and loss of nutrients through erosion, these soils do not support good vegetation cover. Extraordinary care is needed for realising yield from such lands.

INHERENT FERTILITY STATUS

The arid soils are generally low in organic carbon and nitrogen content, medium in available phosphorus and medium to high in available potassium (Table 2). The sandy soils contain 0.1 per cent organic carbon, 10 to 15 kg ha^{-1} phosphorus and 140 to 200 kg ha^{-1} potassium in available form. With increase in clay and silt content there is increase in the content of these nutrient elements. The clay loam soils contain 0.3 to 0.4 per cent organic carbon, 15 to 20 kg ha^{-1} phosphorus and 200 to 700 kg ha^{-1} potassium available to plants (Joshi, 1989). The available iron, manganese, zinc and copper content of these soils vary widely (Table 3). The contents of copper and manganese are generally more in medium to fine textured soils and that of available iron and zinc are higher in coarse textured soils. Even the dune sandy soils (Joshi and Dhir, 1983) and the salt affected soils (Joshi *et al.*, 1981, 1982) have been found to contain adequate levels of micro-nutrients to meet the requirement of the crops and the natural vegetation growing in the region.

Micronutrients are well distributed in the soil profile (Fig. 1). These are sufficient to meet the requirement of deep rooted shrubs and trees. The applicaction of organic matter, nitrogen, phosphorus and potassium should be made as per the requirement of the crops. Under agroforestry, there is differential depletion of nutrients from soil and therefore, continuous monitoring of the nutrient status of soil is essential.

TABLE 2. Major nutrient content of arid zone soils

Soil group	Organic carbon (%)	Available phosphorus (kg ha^{-1})	Available potassium (kg ha^{-1})
Dune and interdune	0.10-0.16	10-15	116-392
Light brown sandy soils	0.12-0.15	9-15	78-482
Hard pan soils	0.10-0.15	9-15	70-300
Gray brown loam	0.25-0.80	11-39	105-890
Brown light loam	0.15-0.25	15-20	300-890
Salt affected	0.08-0.73	9-25	67-440

TABLE 3. Micro nutrient content (mg kg^{-1}) of arid zone soils

Soil groups	Fe	Mn	Zn	Cu
Dunes and inter dunes soils	3.5-27.5 (7.8)	1.1-16.1 (6.1)	0.27-1.91 (0.80)	0.12- 1.52 (0.44)
Light brown sandy	3.4-8.6 (6.2)	4.0-13.4 (8.0)	0.41-1.32 (0.85)	0.36- 0.89 (0.50)
Hard pan soils	3.4-8.6 (6.2)	4.0-13.4 (8.0)	0.41-1.32 (0.85)	0.36- 0.89 (0.50)
Brown light loam	1.6-16.0 (5.8)	5.4-25.0 (11.2)	0.27-3.28 (9.97)	0.44- 0.93 (0.69)
Gray brown loam	3.0-10.4 (6.2)	2.2-16.0 (8.7)	0.36-2.36 (0.88)	0.36- 1.80 (0.90)
Salt affected	2.6-16.0 (6.07)	6.0-29.0 (12.0)	0.56-3.37 (1.10)	0.55- 2.08 (1.10)

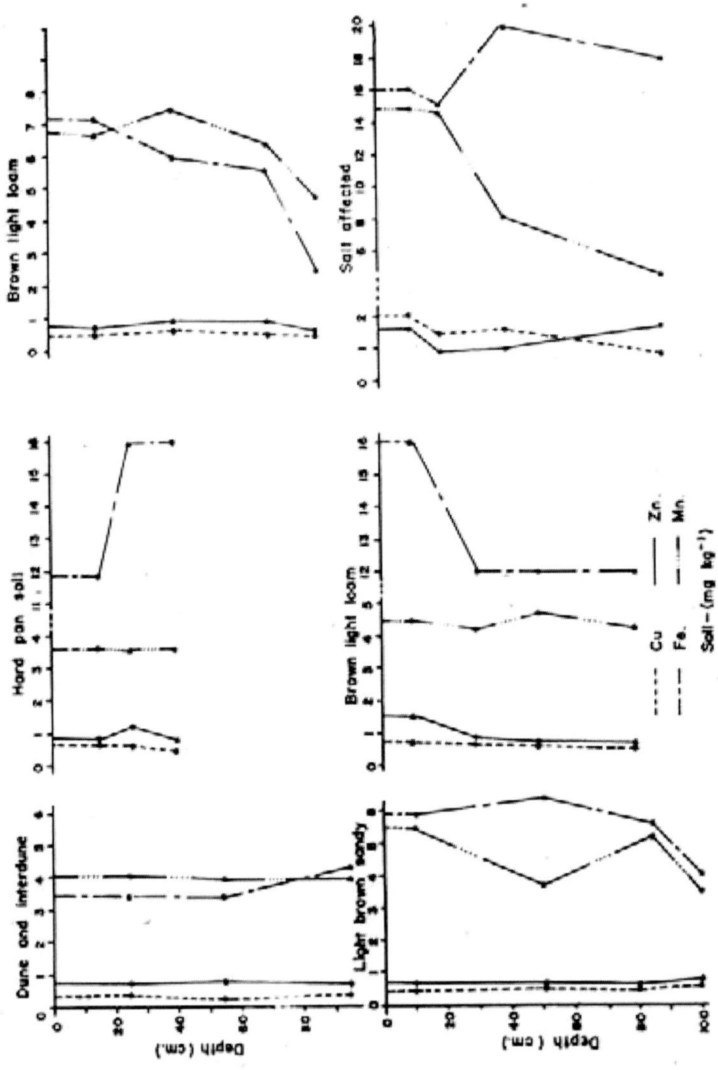

Figure 1. Profile distribution of available micronutrients in arid zone soils

SUITABILITY OF ARID SOILS FOR AGROFORESTRY

Based on inherent soil characteristics, climate and associated land features like topography, slope, erosion, etc. dune and interdune soils, hard pan soils, salt affected soils and shallow gravelly soils qualify for land use capability class VI and VII and are mainly suitable for silvicultural and silvipastoral land uses. However, because of socio-economic constraints, these lands are being cultivated for rainfed crops in *kharif*. The crop yields are meagre and this practice of cultivation inflicts damage to soil.

The slopy and undulating topography of the dunes impose severe restrictions on the agricultural use of land. Ploughing and other soil manipulation practices also set the aeolian hazards. These soils are ideal for growing trees and shrubs, and grasses in the inter row spaces. The interdunal flat lands, however, can be used for cultivation of rainfed crops. Deep sandy soils associated with dune and interdune have low moisture retention capacity and high infiltration rate. The water received through rainfall in good and normal years reaches deep into the soil which can be utilized by the tree component of agroforestry system. These coarse textured soils have non capillary porosity due to which the upward movement of moisture within the profile is limited. Such localised moisture can be utilized by the grass, shrub and tree community. Tree and shrub component of Agroforestry system cover the surface and protect the sandy soils from wind erosion. In addition, the leaves shed by the trees enrich and improve the fertility status of soils.

The available water capacity of shallow gravelly soils is very low and much of rain water either percolate through the loose gravelly mass or drained as runoff. Such areas do not permit growth of crop plants. On such lands bushy shrubs and climatically adopted trees can be grown and in between tree rows, short duration annual grasses may provide some biomass. This will also help in checking the soil erosion due to rain water.

The hard pan soils have a compact horizon below 30 to 40 cm depth which is impermeable to roots and water. Because of thin solum, the stored available moisture is inadequate to meet the requirement of crop plants and in years of low rainfall or during prolonged dry spell the crops fail. For raising trees on such lands the pits should be dug and the impermeable layer broken before planting the sapling. In between the tree rows, grasses/shallow rooted creps can be cultivated.

On salt affected soils, due to high salinity in the profile and low infiltration rate, cropping is not possible unless reclaimed. Agroforestry involving trees and grasses is the only proposition on such lands. For establishment of tree saplings the pits are dug and the saline soil is replaced by good soils. Ridges of 0.8 to 1 m are created in east-west direction and the grass seeds are sown on ridges in between the tree rows. This will also help in improving the infiltration rate of soil and leaching down the salts.

The brown sandy, brown light loam and gray brown loam soils are suitable for agricultural purposes. These soils have higher available water capacity. In the years of good rainfall, particularly on gray brown loam soils, crops like mustard, gram and wheat can be grown under conserved moisture. In the situations of availability of irrigation, bumper crops of wheat, mustard, chillies and other cash crops can be taken. Agroforestry on such soils should be agriculture oriented and tree component should include horticultural plants like *ber, anar,* lemon etc. so that the excess water which percolates beyond the reach of crop plants, can be utilized by these trees and thus generate additional income.

In the rainfed farming situation, particularly on brown sandy soils of sandy plains, agroforestry is the most suitable proposition. Due to frequent droughts failure of rainfed crops is a characteristic feature on these soils. It involves high risk in application of fertilizers and other inputs, because of large variations in the rainfall received during the growing season. The agroforestry involving growing of *khejri (Prosopis cineraria)* and *ber (Zizyphus* sp.) ensures some biomass production as well as economic stability. Improvement in the organic matter content, fertility status of soils and conservation of moisture by providing shade is in addition to the forage available from the foliage of trees/shrubs.

Thus under arid climatic conditions of Rajasthan, the soil resource can be better utilized by adopting agroforestry rather than the monoculture. However, the choice of the plant species and the management techniques need proper understanding of soil characteristics in terms of their limitations and potentialities.

REFERENCES

Dhir, R.P. 1977. Western Rajasthan Soils : Their characteristics and properties. pp. 102-115, In "Desertification and Its Control", ICAR, New Delhi.

Joshi, D.C. and R.P. Dhir. 1983. Distribution of micronutrient forms along duny landscape. Ann. Arid Zone 22 : 135-141.

Joshi, D.C. and R.P. Dhir. 1994. Amelioration and management of soils irrigated with sodic water in the arid region of India. Soil Use and Management 10 : 30-34.

Joshi, D.C., R.P. Dhir and B.S. Gupta. 1981. Forms of iron and manganese in some soils of arid Rajasthan. J. Indian Soc. Soil Sci. 29: 462-468.

Joshi, D.C., R.P. Dhir and B.S. Gupta. 1982. Distribution of different forms of copper and zinc in some soils of arid Rajasthan. J. Indian Soc. Soil Sci. 30 : 547-549.

Joshi, D.C., N. Singh and A.S. Kolarkar. 1988a. Forms of iron and manganese and their availability in salt affected soils of Guhiya catchment. Tropical Agriculture 64 : 41- 48.

Joshi, D.C., N. Singh and A.S. Kolarkar. 1988b. Influence of salinity and associated soil properties on the available zinc and copper in salt affected soils of Guhiya catchment. J.Indian Soc. Soil Sci. 36 : 833-836.

Kolarkar, A.S., R.P. Dhir and N. Singh. 1980. Characteristics and morphogenesis of salt affected soils in south eastern arid Rajasthan. J. Indian Soc. Photo Int. & Remote Sensing 8 : 31-41.

2

Climatic Features and their Modifications through Agroforestry

A.S. Rao

Rainfall, solar radiation, temperature and windspeed conditions of an area mainly determine the suitability of a particular agroforestry system. Besides the day to day weather changes, considerable modifications occur in water and light availability through interactions of these tree/crop systems which have their own biological limits for growth and production. Trees very much protect the soil and crops grown in the interspaces by controlling wind erosion, reducing evaporative demand and protecting the crop from hot, desiccating winds and drifting sand. Before the introduction of any agroforestry systems, their climatic and water requirements should be examined with that of the existing environment for its successful establishment.

RAINFALL AND ITS DISPOSITION

The key weather parameter that determines the success of agroforestry systems in arid and semi-arid regions is rainfall and its distribution. The arid and semi-arid regions of India receive an annual rainfall ranging from less than 150 mm to more than 1000 mm with its coefficient of variation ranging from 30 to 70 per cent. Tree/crop species intercept part of the rain water which is lost through evaporation. Interception losses from trees/crops depend upon (i) rate of evaporation of intercepted water, (ii) canopy storage capacity, and (iii) intensity and its distribution of rainfall. The amount of water which drips from the canopy and falls directly through the gaps is usually called throughfall. The sum of throughfall and stemflow is the net precipitation which goes to the soil. A portion of it is utilized by the trees and inter crops for their actual growth. The interception losses from some trees varied from 0.1 to 1.5 mm (Rao and Vamadevan, 1986; Rao, 1987). The canopy storage capacities were 1.0 to 2.1 mm for coniferous type, 0.4 to 1.0 mm for deciduous type and 0.4 to 2.8 mm for herbaceous type vegetation (Rutter, 1975). Under arid conditions of Jodhpur, the rainfall interception losses from 13 years old *Acacia tortilis* and 7 years old *Holoptelia integrifolia* were 14 to 33 per cent and 3 to 8 per cent, respectively (Ramakrishna and Sastri, 1977; Ramakrishna, 1984). As much as 5.8 mm was intercepted by *Prosopis specigera* in three days without reaching the soil surface (Bhatt and Lahiri, 1964). However, rainfall intercepted by the foliage was directly absorbed by the trees and this led to a significant increase in the relative turgidity of leaves immediately after the shower.

A larger portion of rainfall is lost through transpiration and soil evaporation. Plants often transpire more water than their own body weight in a day and hence the water comes through uptake of soil water and the system is called soil-plant-atmosphere continuum. However, the soil and plants have their own resistance to loss of water under limited moisture conditions. The potential soil evaporation and plant transpiration are governed by vapour pressure deficit of the air, windspeed, heat energy and net radiation available in the system. Under a reasonable dense plant cover the amount of water lost through evaporation from the soil surface is probably small, but it can be substantial from exposed bare soil, especially when irrigated. Under high evaporative conditions, as soil moisture becomes limiting, leaves experience an increasing degree of water stress. Besides a reduction in transpiration, many other physiological processes are adversely affected.

Oduol *et al.* (1986) observed no significant differences in protein content in pods of *Prosopis* due to the climatic differences at two sites, but sugar content in pods across the irrigation treatments differed significantly. Areas with high moisture stress had high sugar in the pods of *Prosopis alba*. The water use efficiency of sorghum (*cv* SPV-96) grown in the interspaces (10x3 m) of *Acacia nilotica* var. *cupressiformis* varied from 0.33 to 2.27 kg mm^{-1} ha^{-1}. Three years old trees did not compete for moisture and light with intercrops viz. sorghum, clusterbean and *moong* (Rao *et al.*, 1993). A study of soil moisture regimes below five desertic trees showed maximum moisture under *Tecomella undulata* followed by *Prosopis cineraria*, *Acacia senegal*, *Albizzia lebbek* and *Prosopis juliflora* (Gupta and Saxena, 1978). As a result, the dry forage yield under *P.*

cineraria was 2.3 t ha^{-1} as against 1.66, 1.32, 0.85, and 0.78 t ha^{-1} obtained under *Tecomella undulata, Albizzia lebbek, Prosopis juliflora* and *Acacia senegal*, respectively (Shankar *et al.*, 1976).

SOLAR RADIATION/LIGHT AVAILABILITY

Solar radiation plays an important role as a regulator and controller in processes of growth and development. More than 70 per cent of solar radiation is absorbed by the plants and converted into heat energy for transpiration and convective heat exchange with the surrounding air. These exchanges determine the temperature of leaves and other plant organs. A portion of solar radiation (about 28% in 0.4 μ to 0.7μ) is used in photosynthesis and stored chemically in high energy organic compounds.

For estimating the potential productivity of crops, growing alone or as mixture, it is necessary to know quantum of light/photosynthetically active radiation (PAR) intercepted. This can be estimated by knowing the light transmission properties of whole units (trees, bushes) and the light intercepted by different layers. Irradiance above a critical level is necessary for some plant processes. In multi-storey systems where the trees form a closed canopy, light transmission to the crop is controlled by vertically-summed LAI of tree and the relevant light extinction coefficient k. In a row and alley systems in which rows or belts of trees are so dense that no crop can grow under them are separated by clear alley-ways in which crops are grown. The light available to the crop is a function of the pattern of shadows casted by the belts of trees which in trun depend on tree height, latitude, row orientation season and time of day. In intermediate agroforestry systems, light is transmitted to the crop by passing the tree altogether.

For many crop plants, the relationship between dry matter yield and intercepted energy is approximately linear (Monteith, 1977) and the fraction (F_{max}) of the PAR intercepted by the crop sets the upper limit for its potential productivity. With continuous- canopy crops, the light interception and distribution is governed by the equation :

$$\frac{I_l}{I_o} = e^{-kl}$$

where I_o is the light intensity above the canopy, I_l the light intensity below a leaf area index of 1 expressed as m^2 leaf per m^2 ground and k is an extinction coefficient calculated after measuring I_o, I_l and 1 under relevant conditions.

There are appreciable changes in the spectral composition of radiation with depth in the canopy, leaves transmitting less of the PAR than the far-red and infra-red parts of the spectrum (Allen *et al.*, 1976). Radiation measurements in 13 years old *A. tortilis* showed that the trees intercepted 70 to 86 per cent of the total radiation resulting in poor growth of understorey *Cenchrus ciliaris* grass though the rainfall during the period was satisfactory (Ramakrishna *et al.*, 1981). These studies indicated that proper lopping of the trees is essential for a successful establishment of silvi-pastoral systems or inter-crops.

It may be preferable to have the entire crop surface irradiated as uniformly as possible or to have a system to keep more shade tolerant crops planted near to the trees. Trees take a long time to grow and to find out the potential number of combinations of tree and crop densities in an agroforestry system is very difficult. Hence a modelling approach to the effects of different arrangements and densities of tree canopies on the light penetrating to the agroforestry floor have several advantages (Jackson and Palmer, 1979).

AIR AND SOIL TEMPERATURES

All plant species grow normally at optimum air and soil temperatures. The photosynthesis-respiration relationship and accumulation of carbohydrates depend upon ambient temperature. Air temperatures beneath a 7 year old *A.tortilis* plantation were lower than those recorded in the open by 0.1° to 0.7°C and 0.6° to 2.0°C during morning and afternoon periods respectively (Ramakrishna and Sastri, 1977). The diurnal variations of temperature in the forests, in general, also decrease from canopy top to the ground surface. Cachan and Duval (1963) recorded a diurnal range of 10.8°C at 46 m height near the tree top whereas 4.4°C at 1 m above the ground surface. Soil temperatures beneath the trees are buffered by the tree canopies. The daily maximum soil temperature beneath the *Acacia* tree cover was lower by 10° to 16°C in the top soil zone (0.5 cm depth) and 4° to 5°C (30 cm depth) than those recorded in the open which indicates a better soil thermal regime under the tree plantation for the adoption of agroforestry system (Ramakrishna and Sastri, 1977).

WIND STRUCTURE AND EFFECT OF SHELTERBELTS

Because of the high water consumption of irrigated crops, especially in arid areas, protection against hot dry winds by shelterbelts is almost essential. Wind influences the transpiration rates of the trees and inter-crops in an agroforestry system. Trees are often used in traditional farming systems of arid regions and as shelterbelts to reduce water losses and increase the productivity per unit of water consumed. The effect of shelterbelt in terms of a reduction in windspeed and evaporation is directly proportional to the height of the belt. However, there is a gradual increase in windspeed with distance behind the belt at a rate determined by its permeability. In common practice, it is defined that a shelter zone is one which reduces windspeed by at least 20 per cent. Depending on windspeed and permeability, the effective shelter distance varies from about 15 to 20 times the height of the belt (W.M.O., 1966).

Three shelterbelts of 300m length with rows of *Acacia tortilis, Cassia siamea* and *Prosopis juliflora* in each belt at 150 m apart were established to study the efficacy of the tree species in arresting the windspeed at 2H, 5H and 10H in the leeward side of the shelterbelt (Gupta *et al.* 1984). Reduction in windspeed was more in the monsoon period (July-September) than in summer period the reduction being maximum at 2H distance from the tree (Table 4).

TABLE 4. Per cent reduction in mean windspeed on the leeward side of shelterbelts

Plant species	Summer season			Monsoon season		
	Distance from shelterbelt			Distance from shelterbelt		
	2H	5H	10H	2H	5H	10H
Prosopis juliflora	33	17	12	38	26	21
Cassia siamea	36	17	13	46	36	24
Acacia tortilis	36	25	13	46	36	20

(Source : Gupta *et al.*, 1984)

Evaporation in arid regions is greatly controlled by windspeed. A decrease of 5 to 14 per cent in pan evaporation values at 2H distance (Table 5) was recorded by Gupta *et al.* (1984). The use of shelterbelts, in general, brought about 50 per cent reduction in the magnitude of wind erosion. In *Cassia siamea* type shelterbelt, the soil loss was 184.3 kg ha^{-1}, while in bare soil (without shelterbelt) it was 546.8 kg ha^{-1}. *Cassia siamea* and *Acacia tortilis* performed better than *Prosopis juliflora* in reducing the windspeed.

TABLE 5. Influence of *Cassia siamea* shelterbelt on pan evaporation

	Pan evaporation (mm day^{-1})			
	April	May	June	July
Windward side	11.0	14.5	13.4	6.0
Leeward side	9.5	13.4	12.4	5.7
Per cent decrease	14	8	8	5

(Source : Gupta *et al.*, 1984)

Some of the tree species suitable for shelterbelts are *Acacia tortilis, Prosopis juliflora, Cassia siamea, Azadirachta indica, Albizzia lebbek, Acacia nilotica* var. *cupressiformis, Tamarix articulata,* etc. (Shankarnarayan *et al.*, 1987).

REFERENCES

Allen, L.H., T.R. Sinclair Jr. and E.R. Lemon. 1976. Radiation micro-climate relationships in multiple cropping systems. In "Multiple Cropping", P.171, ASA Spl. Pub. No. 27, Amm. Soc. Agron., Madison, Wisconsin.

Bhatt, P.N. and A.N. Lahiri. 1964. Some evidence of foliar absorption of water in a xeric tree *Prosopis specigera*. Naturewise 14 : 341-342.

Cachan, P. and J. Duval. 1963. Variations micro- climatiques verticales et Saisonnieres dans la forest Sempervirente de Basse cote d'lveire. Ann. Fac. Sci. Dakar 8: 5- 87.

Gupta, J.P. and S.K. Saxena. 1978. Studies on the monitoring of dynamics of moisture in the soil and the performance of ground flora under desertic communities of tree. Indian J. Ecology 5 : 30-36.

Gupta, J.P., G.G.S.N. Rao, Y.S. Ramakrishna and B.V. Ramana Rao. 1984. Role of shelterbelts in arid zone. Indian Fmg. 24 (7) : 29-31.

Jackson, J.E. and J.W. Palmer. 1979. A simple model of light transmission and interception by discontinuous canopies. Ann. Bot. 44 : 381-383.

Monteith, J.L. 1977. Climate and the efficiency of crop production in Britain. Phil. Trans.Royal Soc., Lond. 281 : 277-294.

Oduol, P.A., P. Felker, C.R. McKinley and C.E. Meier. 1986. Variation among selected *Prosopis* families for pod sugar and pod protein contents. For. Ecol.Mgt. 16 : 423- 431.

Ramakrishna, Y.S. and A.S.R.A.S. Sastri. 1977. Micro-climate under *Acacia tortilis* plantation. Annual Report, CAZRI, Jodhpur, pp. 69-70.

Ramakrishna, Y.S., A.S.R.A.S. Sastri and K.D. Muthana. 1981. A note on the importance of radiant energy penetration in silvipastoral systems. My For. 17 : 1.

Ramakrishna, Y.S. 1984. Forest micro-climate and its importance in agroforestry. In "Agroforestry in Arid and Semi- arid Zones" (K.A. Shankarnarayan, Ed.), CAZRI, Jodhpur, p. 9.

Rao, A.S. and V.K. Vamadevan. 1986. Throughfall, stemflow and interception losses of rainfall from coconut. Mausam 37 : 237-240.

Rao, A.S. 1987. Interception losses of Rainfall from Cashew trees. J.Hydrol. 90 : 293-301.

Rao, A.S., Y.S. Ramakrishna, R.S. Singh and N.K. Chopra. 1993. Water and energy use efficiencies of Sorghum intercropped with *Acacia nilotica* Ann. Arid Zone 32 : 99- 101.

Rutter, A.J. 1975. The hydrological cycle in vegetation. In "Vegetation and the Atmosphere" Vol. I (J.L.Monteith, Ed.), Acad. Press, London pp. 111-154.

Shankarnarayan, K.A., L.N. Harsh and S. Kathju. 1987. Agroforestry in arid zones of India. Agroforestry Systems **5** : 69-88.

Shankar, Vinod, N.K. Dadhich, and S.K. Saxena. 1976. Effect of Khejri tree (*P. cineraria* Macbride) on the productivity of range grasses growing in its vicinity. Forage Res. 2 : 91-96.

W.M.O. 1964. Windbreaks and shelterbelts. Tech. Note No. 59. World Meteorological Organisation, Geneva, Switzerland.

3

Traditional Agroforestry Systems in Western Rajasthan

S.K. Saxena

Western Rajasthan experiences frequent droughts and famines beside overall climatic constraints. In order to evade or minimize adverse conditions the desert dwellers have evolved some agricultural practices to boost the crop production and other related materials like fodder, fuel, fruit and timber. Their centuries-old experiences, passed on from generation to generation, have established a well contained system which accrue maximum benefits by the integration of woody perennials in the mixed farming system. This system in the present context is called agroforestry system.

Their engenuity, according to climatic and edaphic features have selected various suitable drought hardy and multiple use species of trees or shrubs for sustainable production under this system. In normal years of rainfall this system provide sustainable crop and fodder production while under adverse conditions they are able to harvest top feed, fuel to some extent and fruits. Thus adoption of the above strategy at least enable them to evade the drought periods.

Realizing the importance of woody perennials in conserving soil and increase in fertility, the farmers encourage the natural regeneration and establishment of desirable trees and shrubs in their fields. In subsequent years the farmer regulate their density according to the suitability of ploughing. These traditional desert farming systems have actually evolved prior to the fossil fuel utilization for agriculture purposes. Presently, agroforestry has been conceived as the sustainable land management system for practicising of agriculture, forestry and animal husbandry on a long term basis for the benefit of local population in improving the economy and the environment conditions.

TRADITIONAL AGROFORESTRY SYSTEMS

Malhotra (1984) distinguished seven agroforestry zones in western Rajasthan. Based on trees and shrubs species composition, mean annual rainfall and soil types in the western Rajasthan, nine prevalent agroforestry systems were distinguished by the author. By and large, there are three distinct vegetation zones in western Rajasthan.

Shrub Lands

Preponderance of shrub species. It comprises three agroforestry zones. The rainfall zone of 100 to 200 mm covering Jaisalmer, western part of Bikaner, Ganganagar and north western part of Barmer districts.

Wood Lands

Pali, Nagaur, Jhunjhunu, part of Churu, Jodhpur and Ganganagar districts receiving 350 to 450 mm of rainfall.

Combined Zone

The rainfall zone of 200 to 350 mm largely has combination of trees and shrubs. Two distinct agroforestry zones are prevalent.

TABLE 6. Traditional Agroforestry Zones in Western Rajasthan

	Zone	Rainfall (mm)	Habitat	Districts
A.	**Shrub land**			
	Calligonum polygonoides-Halo xylon salicornicum	100-150	Sand dunes, Interdunes	West of Bikaner Jaisalmer,North -west of Barmer
	Ziziphus nummularia-Capp-aris decidua	150-250	Rocky,gravelly pediments & buried pediments	North-west of Jodhpur,Central rocky plateau of Jaisalmer
	Calotropis procera-Calligonum-Clerod endrum phlomoides	200-250	Sandy undulating plains	Barmer (Baytu-Sindri area)

Continue.......

B.	**Trees and Shrubs**			
	Prosopis cineraria-Ziziphus-capparis	250-350	Alluvial plains with kanker pan at 80-150 cm	Jodhpur,Nagaur, Bikaner, part of Barmer
	Salvadora oleoides-Prosopis-Capparis	250-350	Alluvial plains with heavy soils but moderately saline	Pali, Jalor
C.	**Woodland**			
	Prosopis cineraria-Tecomella undulata	250-350	Sandy undulating plains and interdunes	Barmer,Sikar, Jhunjhunu, Churu
	Prosopis cineraria	300-450	Alluvial plains (unirrigated)	Sikar,Jhunjhunu, Nagaur (North & East)
	,*Prosopis cineraria-Acacia nilotica*	350-400	Alluvial plains (irrigated)	Jodhpur, Nagaur, Sikar, Jhunjhunu
	Acacia nilotica A.cupressiformis	400-450	Western foothills of Aravalli range, Alluvial plains	Pali

PREVAILING AGROFORESTRY SYSTEMS IN WESTERN RAJASTHAN

Calligonum-Haloxylon-Leptadenia Zone

This zone covers the western and north-western parts of Jaisalmer and Barmer, western part of Bikaner and Ganganagar districts. Here the average annual rainfall is 100 to 150 mm. The tract is quite dry and highly infested with large dunes. The flanks of the dunes are mostly covered with fresh sand. Many a times it converts into barchan dunes and that too in chain. The activity of fresh sand do not allow any vegetation to establish as a result majority of such dunes are bare, some annual species may come up during monsoon.

Silvi-pastoral system : The tract being deficient in rainfall, only the naturally growing shurbs are maintained. The area is generally known as pastrol. Shrubs and under-shrubs like *Calligonum polygonoides, Leptadenia pyrotechnica, Clerodendrum phlomoides, Calotropis procera, Haloxylon salicornicum, Aerva persica* and *A pseudotomentosa* occur. In Jaisalmer sector, it is *Calligonum* and/or *Acacia jacquemontii* and *Calligonum* while the Barmer sector has its distinction with *Calotropis procera, Clerodendrum phlomoides,* associated with *Calligonum, Lasiurus sindicus.*

Lasiurus sindicus and *Panicum turgidum* are the perennial grasses growing along with the above mentioned species. Due to low and erratic rainfall the tract is largely utilized for animal grazing. *Kharif* crops are cultivated in a good rainfall year and hence once in four to five years the agri- silvi-pastoral system is followed. Camel, Ass and Bullock ploughing is common. *Moth, guar* and local variety of bajra which sustain prolonged span of drought are grown. The crop fields could be witnessed with sown variety alongwith the existing bushes and grass clumps.

Ziziphus nummularia—Capparis decidua Zone

Gravelly and rocky plains, exposed and buried pediments and rocky plateau of Jaisalmer area constitute this zone including Jodhpur district joining Jaisalmer. Gravelly plains of Jodhpur and Nagaur districts also have similar vegetation. Rocky plateau of Jaisalmer, by and large, support much widely scattered shrubs like *Leptadenia pyrotechnica, Ziziphus nummularia, Calotropis procera, Euphorbia caducifolia, Commiphora wightii* and *Capparis decidua*. The drainage lines of this plateau support large bushes of *Salvadora oleoides, Maytenus emarginatus, Lycium barbarum, E. caducifolia, C. procera, L. pyrotechnica, Z. mauritiana, Z. nummularia.*

Silvi-pastoral system : Gravelly plains have good stand of *Ziziphus nummularia* and *Capparis decidua* with varying density (250 to 350 plants ha^{-1}). Many common village grazing lands also support similar vegetation with clumps of grass like *L.sindicus* and serve as a silvi-pastoral system. Where as few pockets with these two main shrub species with some soil accumulation are cultivated for *kharif* crops like cluster bean (*guar*), *moth* and pearl millet (*bajra*). Here crop yields are quite low but the farmers are compelled to grow grain crops. In this tract generally *Ziziphus* shrubs are not harvested. The gradual fall of leaves from December to April provide fodder for grazing animals. Here such patches represent agri-silvi-pastoral system.

Calotropis procera-Calligonum polygonoides-Clerodendrum phlomoides Zone

The sandy undulating interdunal plains of low rainfall zone particularly in Barmer district i.e. Sindri-Baytu tract, a few patches of Phalodi and Shergarh in Jodhpur district, support largely these shrub components. Here *Calotropis* is the dominant species with density of 200 to 400 plants ha^{-1}. This density remains undisturbed when fields are left as long fallow. Before cultivation, majority of *Calotropis* shrubs are cut upto the ground level, or many of them are given the shape of *Calotropis* trees, a few causing hindrance in the cultivation are totally removed. Other associated species growing are mainly *Calligonum* and *Clerodendrum*. At many places *Lasiurus sindicus* and/or *Panicum turgidum* exhibit good stand. During long fallows these tracts largely serve as grazing ground when the farmers follow silvi-pastoral system.

Agri-silvi-pastoral system : *Kharif* crops (once in three or four years, depending upon rainfall) like cluster bean, moth bean, and pearl millet are grown. In general, the crop yields are low but farmers grow them for their immediate requirement. In some fields the density of *Prosopis-Tecomella* or both is 1 to 5 trees ha^{-1}.

Economic benefits : During scarcity and famine periods the *Calotropis* leaves are cut and allowed to dry in the field which can be utilized by sheep, goat and cattle. One year old stems of *Calotropis* are most suitable for making *Jhonpas*. Other sand loving species like *Crotalaria burhia, Aerva persica, A. pseudotomentosa* and *Leptadenia pyrotechnica* are also utilized in thatching *Jhonpas* or huts. Stem fibre, extracted from dry stalks of *Calotropis*, is converted into fine rope for cots. Stands of *Lasiurus* and *Panicum* also provide good forage production. During famine *Aerva* and *Crotalaria* are mixed with normal forage feed.

Prosopis cineraria-Ziziphus nummularia-Capparis decidua Zone

Agri-silviculture : The alluvial plains in the districts of Jodhpur, Barmer, Nagaur and parts of Pali represent this zone. There are 15 to 40 *Prosopis*, 120 to 250 *Ziziphus* and 20 to 35 *Capparis decidua* plants per hectare. Farmers maintain the first two species in the field, while *Capparis* is maintained on the field boundaries. It is difficult to get the impact of *Ziziphus* because it is cut every year to the ground level. Its presence is felt only after the harvest of *kharif* crop. This provides a two-storey system i.e. one of the tree and the other of the shrubs. The association of these species is largely for the livestock and to some extent for farmers' own use. Here the cumulative return form the crop, fodder, fuel, and fruit are quite high, while the total over all input is low. This combination covers the largest area of arid western Rajasthan. In arid tract, the mixture of these three potential species is considered ideal as all the three species are of high economic value. The *kharif* crops grown are : pearl millet, moong bean, moth bean, sesame and cluster bean.

Economic benefits : The vegetative growth of *Ziziphus* by coppicing and root suckers pick up growth during monsoon period. *Ziziphus* plants are harvested by the end of November. They are heaped for 5 to 7 days and then threshed to remove the air dried leaves (pala) for storage. About 200 to 250 kg ha^{-1} of pala production is expected in a normal rainfall year. Recently the introduction of tractor and the use of disc plough has virtually reduced *Ziziphus* population to bare minimum while in majority of cases completely removed. The cultivator feel the pinch and wish to have it again in such a fashion that ploughing with tractor does not get a hurdle.

The trees of *Prosopis* are lopped during the month of December and January to get dry leaves (*loong*) for storage. The lopped branches are heaped for 8 to 10 days and then leaves are removed. Small twigs/branches are used for fencing while larger ones are used as fuel/timber. *Prosopis* trees, sprout back in March and continue to put up new leaves till the onset of monsoon. In case of drought year green loong is harvested and given to animals.

This tree boost the growth of crops under its canopy. Studies of Aggarwal *et al.* (1976) and Shanker et al. (1976) have confirmed that crops/grasses grown under it provide higher biomass production. Gupta and Saxena (1978) observed that *Tecomella undulata* had the highest soil profile moisture followed by *P. cineraria*.

Salvadora-Prosopis-Capparis Zone

Flat alluvial plains of district Jalor and Pali with medium heavy to heavy, moderately saline soils comprise this zone. Climatically this zone has been calssified under semi-arid zone having 350 to 450 mm annual rainfall zone. Here the water table is 10 to 12 m deep and the water is either brackish or saline. The tract has both irrigated and unirrigated areas. Good density of *Salvadora* trees (10 to 20 ha^{-1}) occur in unirrigated areas but decrease under irrigated conditions (5 to 10 trees ha^{-1}). In the past there used to be good density of *Salvadora- Prosopis* in the fields but with the introduction of tractor the farmers are compelled to remove many of the existing trees. However, *Prosopis* trees are least distrubed . Secondly, their natural occurrence is also quite sparse (5 to 10 trees ha^{-1}). Shrubs of *Capparis or Ziziphus*, as the main component of this system, are generally maintained on the field boundaries.

Agri-silviculture : Rainfed crops grown during *kharif* include green gram (*moong*), moth bean, cluster bean (*guar*), sesame (*til*), and pearl millet (*bajra*). Fallow system is also practised. Crops grown in irrigated areas during *rabi* include wheat, barley, cumin and mustard (*raida*).

Economic benefits : Large scale collection of *kair* fruits (*Capparis*) is done for domestic consumption. Plant species like

Acacia nilotica, var. *cupressiformis* and *Z.mauritiana* provide top feed (foliage) to sheep and goats during lean period. Ripe pods of *P.juliflora* become permanent proteinaceous diet to the grazing animals. *Salvadora* fruits (*pillu*) are harvested and used by local population. Its' seeds contain 45 per cent non-edible oil and hence the dried fruits are fed to milch animals to obtain milk of high fat content. Presently seed is marketed to get extra revenue.

Khejri tree is generally not lopped. The green twigs are directly fed to sheep and goats during summer months. Many of the irrigated fields in Jalore have pomegranate orchards. In early stage of establishment leguminous crops are taken and at later stages only grass components are either harvested or grazed by animals once the fruit crop is over. Hence, this zone practices agri-horticulture, agri-horti-pastoral and agri- silviculture systems.

Prosopis cineraria-Tecomella undulata Zone

In western Rajasthan, two distinct zones of *Prosopis- Tecomella*, one in 200 mm rainfall zone of Barmer district and the other in more than 350 mm rainfall zone tract of Lachmangarh and Fatehpur tehsil of Sikar district may be identified. The interdunes with deep to very deep sand deposition in interdune of Chohtan tehsil are well wooded with this species (as high as 60 trees ha^{-1}). In recent years this tract has experienced complete removal of old trees of *Tecomella* for industrial and export use. However, one can witness about 100 to 140 trees ha^{-1} coppice shoots, developed from root suckers.

Agri-silviculture system : In both the areas agri- silviculture system is followed. *Kharif* crops like pearl millet (*bajra*), green gram (*moong*), moth bean and cluster bean (*guar*) are grown. Here *Prosopis* is generally lopped, whereas *Tecomella* is left as such. The leaf fall from both the tree species enrich the fertility status of the soil (Aggarwal *et al.*, 1976).

Economic benefits : *Tecomella* being a timber plant is not disturbed even if some ploughing or other difficulties are faced by the farmers. Studies carried out by Shankar *et al.* (1976) on forage production under various trees including *Tecomella* have shown that it stands second to *Prosopis* in gross biomass production (1.66 t ha^{-1}). In Barmer, cowpea and cluster bean (*guar*) are invariably grown in association with *Tecomella*. Flowers and leaves of *Tecomella* are relished by wild life (Bohra *et al.*, 1992). Goat and Camel also consume its flowers.

Prosopis cineraria Zone

Central, south and western parts of Sikar district, south and eastern part of Nagaur district represent this zone. The soils are loamy sand to sandy and very deep (3 to 8 m). The average rainfall received in this tract is 450 mm. One can see the optimum expression of *Prosopis cineraria* in this zone. Whole tract is well wooded with *P.cineraria* stand with 95 per cent relative density in the cultivated area.

Agri-silviculture system : By and large, this system is followed by the cultivator of this zone. Due to poor quality and quantity of water only *kharif* crops like pearl millet (*bajra*), cluster bean (*guar*), green gram (*moong*), sesame (*til*), cowpea and rarely sorghum (*jowar*) are grown. The density of trees has no adverse effect on the crop production. Opening of tree canopy during monsoon, due to lopping, provide desirable canopy cover which is beneficial to the growing crops. High tree density (50 to 120 trees ha^{-1}) in no way have any competition with shallow rooted *kharif* crops. Besides, the gradual leaf fall (September to December) before lopping, enriches the fertility status of the soil. During cropping period the growing crops do not experience any wind erosion hazard. The local farmers has a feeling that overall ecology of the region gets modified with *Prosopis* woodland and better crop yields are obtained.

Economic benefits : *Prosopis* trees of this region are quite healthy representing good crown cover, girth, and height (42 to 56 m^2, 120 to 150 cm and 8 to 14 m, respectively). These stand provide high *loong* production. On an average 1.5 t ha^{-1} of air dry loong (20 kg $plant^{-1}$ with 80 trees ha^{-1}) and fuel wood are obtained. As per the prevalent rates of *loong* and fire wood one may earn Rs. 3000 per hectare, in addition to the crop harvest. The other associated species include *Tecomella undulata*, *Ziziphus mauritiana*, *Z.nummularia* and *Capparis decidua*. All these species constitute 5 per cent relative dominance. Some fields have also been found with *Saccharum bengalense* (*munja*), which is also an economic plant. However, its density need regulation.

Prosopis cineraria-Acacia nilotica Zone

Semi-arid areas of western Rajasthan covering district Pali, Jalore, Sikar, Jhunjhunu and arid areas of district Ganganagar having good facility of canal irrigation are well studded with healthy plants of *P.cineraria* and *A.nilotica* in the cultivated fields. This zone largely constitute the flat older alluvial plains with fairly good quality and quantity of water. The soils are sandy loam to sandy clay loam, deep to very deep. It is reported that in Ganganagar district there was very good stand of *Prosopis* alone before the introduction of canal in 1930. The trees has now drastically decreased due to tractor ploughing. At present the tree population is restricted to 20 to 25 plants ha^{-1} which include *A. nilotica* as well. In Sikar district, the density may go as high as 35 plants ha^{-1}. Under irrigated conditions, the trees of both the species are healthy with clean bole, large crown (42 to 64 m^2) and good girth (120 to 175 cm).

Agri-silviculture : Before taking *rabi* crops, the trees are lopped in October to open the canopy. Once in 3 to 5 years, the crown spread is also reduced by cutting the old branches. This provides wood to the cultivator. Some cultivators do not cut *Acacia* species. Both the trees are deep rooted and do not exert any competition to growing crops. Crops like wheat, barley, mustard, gram, etc., are taken during *rabi* season, while in *kharif*, cotton, sorghum, pearl millet and sesame are grown.

Economic benefits : Farmers in Sikar and Jhunjhunu districts grow *Alianthus excelsa* around the field boundaries. Leaves of this plant are relished by sheep and goat. Annually *Alianthus* plants are sold at the rate of Rs. 35 to 40 per plant for its foliage. In this agri-silviculture system, the farmers raise 80 to 100 cm high field bunds on which live hedges of *Saccharum bengalense, Ipomoea carnea, Opuntia* sp., *Agave sisalana, Ziziphus nummularia* and *Acacia senegal* are raised. In case of *Saccharum*, some revenue is earned while the individual household needs material for hut thatching. These live hedges also act as wind breaks and create favourable micro-climate for crops, beside protecting crop from the stray animals. On an average a farmer earns an additional revenue of Rs.2500 ha^{-1} from sale of loong, pala, munja and fire wood from the trees. The live hedge/fence itself makes a subsystem of agroforestry.

Acacia nilotica-A. cupressiformis Zone

This zone lies along the Aravalli hill ranges in the east of Pali district. It consists of flat older alluvial plains and buried pediments with moderately heavy and deep soils.

Agri-silviculture system : In this zone limited irrigation facilities are available. The fields have the dominance of *Acacia nilotica* and *A.cupressiformis* with 15 to 35 plants ha^{-1}. They are maintained by the farmers in such a way that cultural operations by the tractor are without any hindrance. Farmers encourage a higher density of *A.cupressiformis* (50 plants ha^{-1}) in their fields. The restricted pyramidal crown of this species does not impose any shade to the growing crops. It is supposed to be the best species for agroforestry in the irrigated area and this should be encouraged. Crops grown during *rabi* season are wheat, mustard and gram, whereas the *kharif* crops include maize, sorghum, mung bean, cotton and sesame.

Economic benefits : Associated trees of this agroforestry system in cultivated fields are : *A.leucophloea, P.cineraria, Z.mauritiana, Alianthus excelsa* and *Salvadora oleoides*. All these species also provide top feed for the livestock. These species are generally maintained around the field boundaries. Direct lopping of green twigs is done to feed the sheep and goat. Pods of both the *Acacia* species are also fed to cattle, sheep and goat. Small lopped twigs are used as fencing material. Moderately thick branches serve as fuel wood. Good quality wood of *Acacia* can be obtained in ten years which can be utilized for agricultural implements and household furniture etc. Indirectly the animal upkeep and their regular fodder and top feed supply is made available from the valuable trees of multipurpose.

CONCLUSIONS

Existing agroforestry systems followed in the arid areas are efficient and productive.

Plant species adopted in various agroforestry systems are more compatible with annual crops in irrigated or rainfed cropping.

Some species enrich the soil fertility and provide nutrient to the crops i.e. *P.cineraria*.

Various agroforestry systems provide different economic returns. More studies are needed to catalogue this information.

The adopted species in many ways check environmental degradation, especially during consecutive droughts and famine conditions.

REFERENCES

Aggarwal, R.K., J.P. Gupta, S.K. Sexena, and K.D. Muthana. 1976. Studies on soil physico-chemical and ecological changes under twelve years old five desert tree species of western Rajasthan. Indian For. 102 : 863-872.

Bohra, H.C., S.P. Goyal, P.K. Ghosh and Ishwar Prakash. 1992. Studies on the ethology and ecophysiology of the antelopes of the Indian desert. Ann. Arid Zone 31 : 83-96.

Gupta, J.P. and S.K. Saxena. 1978. Studies on the monitoring of the dynamics of moistrure in the soils and the performance of the ground flora under desertic communities of trees. Indian J. Eco. 5 : 30-36.

Malhotra, S.P. 1984. Traditional agroforestry practices on arid zone of Rajasthan. In "Agroforestry in Arid and Seim-arid Zones" (K.A. Shankarnarayan, Ed.), pp. 263-266.

Shankar, Vinod, N.K. Dadhich and S.K. Saxena. 1976. Effect of Khejri tree (*P.cineraria* Macbride) on the productivity of range grasses growing in its vicinity. Forage Res. 2 : 91-96.

4

Some Alternative Production Systems and their Management for Sustainability

J.P. Gupta

Crop production in about 60 to 70 per cent of cultivated area in India and 95 per cent in western Rajasthan depends upon monsoon rains. The mean annual rainfall in arid regions is not only low, ranging from 250 to 300 mm, but highly erratic in distribution thereby leading to low yields and frequent failures of crops. Therefore, it is not appropriate to depend completely on traditional agriculture which is uncertain and extremely risky but to develop alternative land use systems which are more sustainable, meet the fuel wood and fodder needs,stabilize the income; and at the same time protect the environment. Successful establishment of such systems could be possible with the use of suitable management practices and conservation measures.

TRADITIONAL AGROFORESTRY APPROACHES

Arid zone farmers have been following agroforestry since centuries by allowing to grow and protecting *'khejri'*, *'ber'* and *'babool'* trees in the cultivated fields. There is a common belief that crop productivity increases in association with *khejri* trees. In a survey on agroforestry systems, Gupta and Sharma (1989) observed density of *Prosopis cineraria* (*khejri*) 30 to 40 trees ha^{-1} in Jodhpur and Nagaur regions having an average annual rainfall of 300 to 500 mm (Table 7). It increased to as high as 104 trees ha^{-1} in Sikar region with a rainfall of 450 mm. Though the growth of trees in terms of height, canopy diameter and collar diameter was better in Jodhpur region, the total canopy area was more in Sikar region because of higher tree density. Pearl millet, mung bean, moth bean, cluster bean and sesame crops are mainly grown in these regions. Because of uncertainty of rainfall no fertilizer is given though manured @ 1 to 2 tons ha^{-1}. Tree density higher than 56 trees ha^{-1} showed reduction in the yield of pearl millet in Sikar area inspite of higher rainfall (450 mm). Crop yields, from below and away from the canopy, were generally comparable showing thereby, no adverse effect of tree canopy on crops.

TABLE 7. Effect of rainfall and *khejri* density on the yield and dry matter production of some arable crops

Location	Mean rainfall (mm)	Tree density ha^{-1}	Crops	Crop yield kg ha^{-1}			
				Crop below tree canopy		Away from tree canopy	
				Grain	Straw	Grain	Straw
Banthri (Nagaur)	332	28	Til	320	163	320	1320
Bugarda (Nagaur)	333	32	Bajra	340	133	210	1720
Tarnau (Nagaur)	333	48	Guar	200	261	880	3800
Sujangarh (Churu)	368	40	Bajra	720	128	850	1650
Nettra (Jodhpur)	373	48	Bajra	800	124	580	2920
Kachawa (Sikar)	470	56	Guar	080	67	110	1130
Antroli (Sikar)	470	56	Bajra	400	110	180	960

IMPROVED AGROFORESTRY SYSTEMS

Although *khejri* with field crops is a naturally existing agroforestry system in the arid areas, it has a limitation of slow growth and therefore availability of fodder and fuel from this tree takes a long time. There are other multipurpose trees which have compatibility with crops and grasses in arid regions and can provide early returns.

Agri-silviculture system

Tree species like *Acacia nilotica, Acacia senegal, Tecomella undulata, Hardwickia binata, D. nutans* and *C.mopane* could be useful components for agroforestry systems in arid regions. Barring *Tecomella undulata* and *Acacia senegal* (which are slow growing but have economic value), others are relatively fast growing and have fodder and fuel value and thus could be used for agroforestry purposes. The grain yields of mung bean and cluster bean crops grown under eight year old unlopped (*Holoptelia integrifolia*) trees were lower than that in the open field but were at par when the trees were lopped (Muthana et al., 1977). Although *Acacia albida* reduced the yield of 'guar' and 'moong' bean crops after three years of growth but *Prosopis cineraria* did not (Shankarnarayan, 1984). Sharma et al.(1984) observed no significant differences between the yields of *guar*, pearl millet, *moong* and moth crops grown in accociation with *Ziziphus rotundifolia* and open field.

Agri-horticulture system

Ziziphus rotundifolia is an important fruit tree of the arid areas. *Z. nummularia* (Jhar ber) grows naturally at the farmer's fields and does not compete with the crops. However, budded *ber* (*Z.mauritiana*) has also been found to grow very well under arid conditions when planted either with water harvesting system or with supplemental irrigation. Three year old plantation of *ber* at a density of 400 plants ha^{-1} in association with mung beans performed very well in 1988 with a seasonal rainfall of 210 mm only (Table 8).

TABLE 8. *Moong* yield as affected by intercropping with *ber* plants

Treatment	Rainfall (mm)	Ber yield (kg ha^{-1})	Grain yield (kg ha^{-1})	Economic return (Rs ha^{-1})	Net profit (Rs ha^{-1})
Normal planting	210	-	620	4800	-
Intercropped with ber	210	800	160	7680	2880

Intercropped *moong* yielded only 160 kg ha^{-1} as against 620 kg ha^{-1} from pure crop. The fruit yield from the intercropped system increased the net profit to Rs. 2886/-. This shows that this system minimizes the risks in dryland agriculture and helps in imparting economic stability. The system, however, is generally recommended for areas with rainfall more than 250 to 300 mm.

Horti-pastoral system

Like *P.cineraria*, *Ziziphus nummularia* is another important top feed species. Even in drought years when crop fail, the top feed of this species come to the rescue of farmer's animals. In recent past, the intensive use of tractor has wiped out the deep rooted bush from the farmer's field. The results of more than eight years studies (Sharma and Vashishta, 1985) in sandy range lands at Samdri (western Rajasthan) have shown that plantation of two jujube species viz. *Ziziphus rotundifolia* each with four densities i.e. 280, 140 and 70 plants ha^{-1} in *Cenchrus ciliaris* pasture gave forage yield varying from 624 kg to 844 kg ha^{-1} in different treatments. Different population densities of jujube in the pasture had no significant effect on the yield of pasture indicating thereby that even 280 trees ha^{-1} can safely be planted. *Z.nummularia* started top feed and bush wood production after 4 years.Sharma and Diwakar (1989) worked out the economics of horti-pastoral system and found the return to be remunerative even with the higher discount rate of interest (20%) and increased investment cost (25%) due to hike in prices of the inputs.

Silvi-pastoral System

This system is essentially for the degraded lands in areas having rainfall less than 200 mm. In this system the grass is intercropped with fuel and fodder trees, preferably with those having relatively less water requirement. *Acacia tortilis, Ziziphus rotundifolia, Ziziphus nummularia, Acacia senegal, Prosopis juliflora, C.mopane, D.nutans, H.binata* are some of the suitable trees species for this system.

The promising grasses suitable for low rainfall conditions are : *Cenchrus ciliaris, C.setigerus, Lasiurus sindicus, Panicum antidotale* and *Dichanthium annulatum*. *Lasiurus sindicus* has been found to thrive very well even under extreme arid climate of Jaisalmer which is known as its natural habitat. Studies (Muthana et al., 1985) with *Acacia tortilis* and grass combination from 1977 to 1984 indicated that *Acacia tortilis* at 10m x 5m spacing increased the grass production over the control (pure pasture) particularly during normal and high rainfall years. During low rainfall years, however, the yields were either comparable or marginally decreased. Other tree species such as *Eucalyptus camaldulensis* and *Prosopis cineraria* were also found to perform well. The forage production from the year 1979 to 1981 under these species was comparable with the control (without trees). Studies with 10 years old plantation revealed maximum survivability of *A.tortilis* and

C.mopane (94 to 100%), though the height attained by *A. tortilis* and *A. indica* was higher. The mean forage yield of *Cenchrus* sp. was higher under *D.nutan* and *C.mopane* but the forage yields of *C.ciliaris* and L.sindicus in association with *Z.rotundifolia* were statistically at par with that in the open field (Sharma et al., 1993).

MANAGEMENT PRACTICES FOR IMPROVED TREE CROP PRODUCTION

In order to minimise the risks of crop failures and make agroforestry a success following soil and moisture conservation measures need to be adopted.

Water Harvesting

Water harvesting is an age old practice prevalent in arid Rajasthân. The rain water from catchment is collected in ponds in low lying areas and depressions for drinking, crop cultivation and ground water recharge. Various methods like interplot, inter-row and micro-catchments are artificially made to make more water available for the success of the crop. Micro- catchments (50 cm wide feeding 50 cm furrows) helped in increasing the production of bajra and *guar* crops even during drought years (Table 9). The technique of compacted circular catchment (1.5 m radius with 5 per cent slope) was highly useful in increasing the profile water storage and establishment of plants (Gupta and Muthana, 1985). Intergrated water harvesting involving catchment construction, application of sub-surface barrier and manures could markedly improve the growth and yield of even two years old *Ziziphus mauritiana* plants grown in arid areas (Table 10). Application of minimal quantities of some water absorbants in the planting pits further helps in better establishment and growth of plants (Gupta and Sharma, 1991).

TABLE 9. Effect of water harvesting on the yield of crops

Treatment	Grain yield (kg ha^{-1})	
	(1985)	(1989)
Control	770	570
Micro-water harvesting	970	630

TABLE 10. Effect of integrated water harvesting (IWH) on the growht of '*ber*' Plants (3 Yrs old) in 1987

Treatments	Fruit yield (3 yrs)	Plant height (cm)	Canopy diameter (cm)	Branch number per plant	Dry wt. of pruned material (g/plant)	Frost affected plants
Control	288	85	197	30	77	3
IWH	591	119	252	42	508	1

Tillage

Tillage is an essential operation both for trees and crops especially for seed bed preparation, weed removal, increasing water infiltration and storage etc. However, excessive tillage may destroy the soil structure and make the soil more vulnerable to wind and water erosion. The results of a study on a loamy sand soil show (Table 11) that soil preparation with minimum tillage (one disking) before sowing of *kharif* crop is adequate for minimising the weed population, improving the soil physical conditions; and growth and yield of bajra, cowpea and *moong* crops (Gupta, 1986). A post-sowing cultivation 25 days after sowing of bajra checked the weed growth, reduced the crop weed competition and thus markedly increased the yield.

TABLE 11. Effect of tillage on yield of crops on a loamy sand soil

Treatments	Mean grain yield (kg ha^{-1})		
	(1982-85)	(1982-84)	(1985-86)
No tillage	1350	760	470
Minimum tillage	1550	1140	580

Manures and Fertilizers

The use of fertilizers in arid areas is generally low owing to low and uncertain rainfall. However, depending upon the availability 1 to 2 t ha^{-1} of organic manure is applied. Gupta et al. (1983) reported that application of 6 t ha^{-1} manure once in four years in seeded rows spaced at 10 to 20 cm is highly useful in improving the soil moisture retention, water use efficiency and yield of *bajra* (Table 12). Though the higher dose of fertilizer is difficult under the extremely arid conditions, 10 to 20 kg N ha^{-1} at the time of sowing has been found useful. A further dose of 10 to 20 kg N ha^{-1} could be given if the subsequent rainfall conditions are favourable. Manuring in association with water harvesting further increases the water use efficiency, both for field crops and tree seedlings.

TABLE 12. Effect of deep drilling of manure on the yield of *bajra*

Treatment/rain fall (mm)	Grain yield (kg ha^{-1})					
	1985	1986	1987	1988	1989	Mean
Rainfall (mm)	152	193	32	235	209	
Control	760	120	-	110	370	270
Manure applied at 10 cm depth	960	250	-	440	900	510

Mulching

Mulching is a practice of covering the soil surface with organic or plastic materials for reducing evaporation losses, soil temperature, weed growth, etc. thereby increasing the water use efficiency. Though crop straws, grass, etc. have been found useful, their availability in arid region is limited particularly in view of the increased fuel and fodder requirements. Perennial weeds, however, are abundant and could possibly be used as a mulch which are otherwise burnt before sowing of the crop. Mulching of inter-row crop spacing with perennial weeds @ 3 to 6 tha^{-1} markedly improve the soil environment by delaying in soil drying, reducing the diurnal variations in soil temperature and checking the growth and population of weeds (Gupta, 1985). As a result, there was an increase in the yields of *bajra* and grain legumes even during drought years (Table 13). Mulching the surface of pit after planting tree seedlings has also been found highly useful particularly in checking seedling mortality and boosting the early growth and establishment.

TABLE 13. Effect of mulching on the crop yield

Treatment	Mean grain yield (kg ha^{-1})		
	Bajra (1982-85)	Cowpea (1982-84)	*Moong* (1985-86)
No mulch	1450	820	520
Mulch	1620	1190	580

Plant Population

Under agroforestry systems, population of the crop plants should generally be low to avoid competition and crop failure, and for better survival during the drought period. Results have shown that plant population greater than 6 lakh plants ha^{-1} significantly reduce the growth and yield of *moong* crop with a seasonal rainfall of 235 mm only.

Sheltérbelt Plantation

It is one of most important long term measures to minimize the risk of crop failures. Plantaition of single or multi-rows of trees along the field boundaries protect the crop from hot desiccating winds, reduce the evaporation losses from the soil and improves the environment for better growth and yield of crops even during low rainfall years beside providing fuel wood and fodder. The role of shelterbelt has been discussed in chpater 7 of this book.

CONCLUSIONS

In view of the low and erratic rainfall in arid areas, agroforestry, agri–horticulture and silvi-pasture are better alternatives to sole cropping for medium to low rainfall conditions, respectively. Water harvesting with micro-catchments, minimum tillage, 1 to 2 tha^{-1} organic manures, 10 to 20 kg N ha^{-1} and low plant population are the management practices for sustainable production.

REFERENCES

Gupta, J.P., R.K. Aggarwal, G.N. Gupta and P. Kaul. 1983. Effect of continuous application of farm yard manure and urea on soil properties and production of pearlmillet in western Rajasthan. Indian J. Agric. Sci. 53 : 53-56.

Gupta, J.P. and K.D.Muthana. 1985. Effect of integrated moisture conservation technology on the early growth and establishment of *Acacia tortilis* in the Indian desert. Indian for. 3 : 478-485.

Gupta, J.P. 1985. Mulch is must. Intensive Agric. XXII : 16-18.

Gupta, J.P. and B.M. Sharma. 1989. Annual Report, CAZRI, Jodhpur, p.31.

Gupta, J.P. 1986. Effect of tillage and mulch on soil and the growth and yield of cowpea grown in the arid tropics. Arid Soil Res. and Rehab. 1 : 161-172.

Muthana, K.D., Gian Chand and G.D. Arora. 1977. Studies on the coppicing behaviour of desert fuel cum fodder species. Annual Report, CAZRI, pp. 70-71.

Muthana, K.D., S.K. Sharma and L.N. Harsh. 1985. Study on silvipastoral system in arid zone. My Forest 21 (3) : 233-238.

Shankarnarayan, K.A. 1984. Agroforestry in arid and semi-arid zone. CAZRI, Jodhpur, p. 295.

Sharma, S.K. and B.B. Vashishtha. 1985. Evaluation of jujube buffel grass horti-pastoral system under arid environment. Ann. Arid Zone 24 : 203-309.

Sharma, S.K. and G.D. Diwakar. 1989. Economic evaluation of horticultural system in arid region of western Rajasthan. Indian J. Range Mgt. 10 : 119-122.

Sharma, B.M., J.P. Gupta, S.S. Rathore and R.N. Prasad. 1993. Evaluation of some silvi-pastoral systems under arid conditions of Rajasthan. J. Trop. Forest. 9 : 103-109.

Sharma, B.M., S.S. Rathore and J.P. Gupta. 1994. Compatibility studies on *Acacia tortilis* and *Zizyphus rotundifolia* with some arid zone crops. Indian For. 120 : 423- 429.

5

Some Improved Silvipastoral Systems for Sustainable Production

S.K. Sharma

The arid and semi-arid parts of the country face severe problems of high population, overgrazing and poor land use, resulting in over exploitation of available resources. Considering that livestock husbandry occupies the most important place in the economy of arid regions and that frequent droughts result in loss of livestock wealth owing to the shortage of fodder resources, it is necessary that fodder production is complimented with the raising of fodder tree and shrub species which can provide not only the much needed forage during the scarcity period but also provide shade and shelter to animals.

Silvi-pastoral system is meant for simultaneous production of wood, top feed as well as grasses and fodder crops for rearing of livestock. Every pastoral site presents distinct management problem. A knowledge of grazing and browse habit of livestock helps in developing a suitable silvi-pastoral model for such sites.

SUITABLE GRASSES AND TOP FEED SPECIES

Suitable trees, shrubs and grass species have been identified for different categories of land and agro-climatic conditions. Some important high yielding perennial grasses are : *Cenchrus ciliaris, Cenchrus setigerus, Lasiurus sindicus, Dichanthium annulatum, Sporobolus helvolus* and *Sehima nervosum*. Some of the recommended tree and shrub species for dry tropics are : *Acacia tortilis, Acacia cupressiformis, Albizzia amara* (Lallei), *Hardwickia binata, Prosopis cineraria, Ziziphus nummularia, Ailanthus excelsa, Colophospermum mopane, Leucaena leucocephala, Dichrostachys nutans* and *Sesbania* sp. Most of these species are of fodder and fuel value.

A number of silvicultural and silvi-pastoral studies have been conducted in order to assess the potential fuel and forage production under different management systems. The adoption of suitable technologies will improve the condition and primary productivity of drylands under different land use classes in future developement programme.

UNDERSTOREY FORAGE PRODUCTION IN AFFORESTED AREAS

The ground vegetation in the afforested area differs in botanical composition under different tree species. Growth characteristics of most of the perennial and annual grass species in respect of their numbers, density and dry matter production, were much better under *Prosopis cineraria* plantation than other tree species (Aggarwal et al., 1976). Ahuja et al. (1978, 1985) and Sharma et al. (1980) studied forage production under afforested tree species commonly found in arid and semi-arid parts of western Rajasthan. The growth and primary productivity of understorey grasses varied in different years and locations. Forage production was maximum in 1973, a good rainfall year with even distribution, and least in 1974 (drought year). Forage yield in 1975, another good rainfall year, was less than that of 1973 due to uneven distribution of rainfall during August, followed by heavy rains during September to October. Contribution of perennial grass was maximum during 1975 as rainfall extended upto middle of October which provided large span of growth period. During 1973, precipitation period was limited upto the middle of September which restricted the growth period of perennial grasses, though proved ideal for the ephemeral species. The forage production under *P.cineraria, Tecomella undulata* and *Albizzia lebbek* did not differ significantly but under *Acacia senegal* it was very low.

PASTURE IN INTER-ROW SPACES OF TREES

The introduction of different pasture grasses between the inter-spaces of established tree species increase the forage production of the system. Shankarnarayan and Saxena (1976) recorded maximum dry matter yield from *Chrysopogon fulvus* (2.4 t ha^{-1}) followed by *Cenchrus ciliaris* var. 3108 (2.2 t ha^{-1}), *C.ciliaris*, var. *Molopo buffel* (1.7 t ha^{-1}), *Sehima nervosum* (1.7 t ha^{-1}) and *C.setigerus* (1.5 t ha^{-1}).

TREE GRASS INTERACTION AT EARLY ESTABLISHMENT PHASE

In silvi-pastoral studies initiated at CAZRI, Jodhpur during 1977, *Acacia tortilis* seedlings were transplanted at 10 x 5m and 10 x 10m spacing in pasture of *Cenchrus ciliaris*. The growth of *Acacia tortilis* seedlings during initial three years suppressed when raised in combination with *Cenchrus ciliaris*. After three years, plants picked up better growth rate, indicating that the plants developed deeper root system and made the best use of the moisture available at the lower soil strata. The mean annual increment (MAI) of *Acacia tortilis* trees was higher in plots without grass than with grass from initial planting stage upto the third year growth period. After this phase, the trend of tree growth with grass was high due to more moisture stored in lower soil strata (Muthana et al., 1985).

Forage Production

Dry forage yield of *C.ciliaris* obtained under different spacings of *A. tortilis* was statistically non-significant in all the years (Table 14). But significant variations in forage yield were observed during different years. This may be attributed to the variations and distribution pattern of rainfall. Studies over a period of seven years have given indication that the combination of trees with grass would yield better returns over pure tree plantation or pure grassland.

TABLE 14. Dry forage yield (kg ha^{-1}) in silvi-pastoral system during different years

Treatments		1977	1978	1979	1980	1981	1982	1983	1984
Annual rainfall (mm)		353.3	354.9	758.1	232.3	316.6	417.2	512.6	231.2
S_1 (10 x 5m)		1252	668	1358	668	482	458	1537	831
S_2 (10 x 10m)		999	567	904	589	350	433	1564	871
S_0 (pure pasture)		1602	535	1140	712	527	222	991	830
Years	SEm±	90.0	CD (0.05)		283.5				
Spacing	SEm+	85.7	CD		Non-significant				

Effect of Coppicing and Spacing

The spacing of top feed species in a silvi-pastoral system plays an important role in forage production. Close spacing reduces the forage component in the system while canopy of grown up trees adversely affect, the forage yield. This is clear from the following forage yield data recorded from coppiced and non- coppiced plots (Muthana et al., 1977).

TABLE 15. Effect of spacing and coppicing of *Acacia tortilis* on forage yield

Spacing (m)	Yield (kg ha^{-1})	
	Coppiced	Uncoppiced
6.0 x 6.0	2356	595
4.5 x 4.5	348	171
3.0 x 3.0	78	39

Silvi-pastoral System in Ravine Land Reclamation

Degraded ravine lands in arid Rajasthan and Gujarat are vulnerable to soil erosion and fast depletion in their vegetation cover. Recent studies (Chinnamani, 1992) in Mahi ravine lands in Gujarat revealed manifold increase in the forage production under rainfed conditions by using different combinations of top feed tree and grass species (Table 16).

TABLE 16. Dry forage production in ravines

Trees		Grass species	Yield (t ha^{-1})
Leucaena leucocephala	+	*C. ciliaris*	6 to 12
-do-	+	*Chrysopogon fulvus*	5 to 12
-do-	+	*D. annulatum*	7 to 12
Hardwickia binata	+	-do-	5 to 9
Anogeissus sp.	+	-do-	6 to 8
Sesbania sp.	+	-do-	7 to 11
Gliricidia sp.	+	-do-	6 to 10
Acacia nilotica	+	-do-	4 to 8

In denuded ravines, introduction of trees and grasses under intensive care and adopting soil conservation measures like gully control, diversion bund, check dam, contour trenches, and bunds gave good establishment in 2 to 6 years. This was followed by protection from biotic factors and closure. This brought in succession of useful vegetation.

Studies have revealed that protected silvi-pasture cover is best in erosion control than other agricultural crops (Table 17).

TABLE 17. Runoff and soil loss in ravines of Vasad

Particulars	Average runoff (mm)	Soil loss (t ha^{-1} Yr^{-1})
Natural fallow	16.5	0.02
Silvi-pasture	1.0	0.00
Agricultural crops	12.0-25.0	2.20-4.80
Denuded ravines	15-30	16-30

CONCLUSIONS

It may be concluded that silvi-pastoral system is ideal for those degraded waste and ravine lands which are not fit for cultivation and need immediate attention for their reclamation. Also, there is a need to recognize the importance of our native tree and shrub species which are well compatible for silvi- pastoral system.

REFERENCES

Aggarwal, R.K., J.P. Gupta, S.K. Saxena and K.D. Muthana. 1976. Studies on physico-chemical and ecological changes under 12 yrs old five desert tree species of western Rajasthan. Indian For. 102 : 863-872.

Ahuja, L.D., C.M. Verma, S.K. Sharma and T.R. Lamba. 1978. Range management studies on the contribution of ground storey (grass) in afforested areas in arid regions. Ann. Arid Zone 17 : 304-310.

Ahuja, L.D., C.M. Verma, S.K. Sharma and T.R. Lamba. 1985. Contribution of grass component (ground storey) in afforested areas in arid regions. Indian For. 111 : 242-248.

Chinnamani, S. 1992. Silvi-pasture and natural grassland of Mahi ravines and ravines of Gujarat. Range Mgmt. and Agroforestry 13 : 175-181.

Muthana, K.D., Gian Chand and G.D. Arora. 1977. Studies on the cappicing behaviour of desert fuel cum-fodder species. C.A.Z.R.I. Annual Report– pp. 70-71.

Muthana, K.D., S.K. Sharma and L.N. Harsh. 1985. Studies on silvipastoral systems in arid zone. My Forest 21 : 233-238.

Sharma, S.K., C.M. Verma and L.D. Ahuja. 1980. Production of ground storey (grass component) in afforested areas in arid regions of India. Ann. Arid Zone 19 : 283-287.

Shankarnarayan, K.A. and S.K. Saxena. 1976. Performance of grasses in silvipastoral systems. Annual Report, CAZRI, Jodhpur, p. 90.

6

Integrated Farming Systems for Higher and Sustained Productivity in Watersheds of Arid areas

T.K. Bhati

The concept of sustainability is far more challenging in the hot arid regions of India because of their complex and multifarious problems i.e. environmental, biotic and socio-economic. Consequently, ecological balance between human (19 million) and animal (23 million) population on one hand; and plant, water and land resources on the other, has set in. If not checked, this may lead to further accentuation of desertification process. Although traditional land use system like mixed cropping, agroforestry, agri-pastoral, silvi-pastoral and *khadin* cultivation are common, crux of the problem is that the ecosystem has never been viewed in totality. Various land use systems viz. agri-pastoral, agri-horticulture, horti-pastoral and silvi-pastoral have been developed and evaluated under different agroclimatic conditions. Integration of these land use systems with appropriate soil and water conservation measures in farming systems development (FSD) approach can cater the demand for food, fuel, fodder and also cash need of the farmer on sustained basis even from these fragile eco-systems.

IMPROVED DRYLAND FARMING TECHNOLOGY

More than 70 per cent population of arid Rajasthan is practising arable farming on 90 per cent of the cultivable area. The average yield of principal crops, pearl millet (*bajra*). moth bean (*moth*) and mung bean (*moong*) are very low and often the crops fail completely.

Further, intense grazing have made these areas highly vulnerable to wind erosion and consequent loss of soil fertility. However, with proper management of inputs, the crop productivity can be improved upto 300 per cent. The main components of the production technology are as follows :

Efficient Management of Rain water

— *In-situ* runoff conservation through inter-row water harvesting/ridge-furrow cultivation attainable with furrow opener attachment with existing cultivator.

— Inter-plot water harvesting (micro-catchment farming) for improving moisture regime in fruit orchards like *ber*.

— Runoff collection in ponds and recycling for raising nurseries.

— Incorporation of bentonite clay as sub-surface barrier 2-3 cm thick to reduce losses of moisture through deep percolation for supporting the establishment of trees.

— Surface mulching with field vegetation wastes @ 4 to 6 t ha^{-1} immediately after sowing. This helps in reducing evaporation losses during seedling emergence and surface crusting.

— One to two interculture operations with blade hoe, sweep hoe, rotary tine hoe upto 30 to 40 days of crop growth for weed control.

Tillage and Sowing

— Deep ploughing in alternate years for moisture conservation in the soil profile.

— Row spacing : For pearl millet and legume crops, row to row spacing of 45 to 50 cm is recommended. In case of hired cultivators when tines fixed at 22 or 25 cm the row spacing can be maintained by skipping alternate rows.

— Seed bed preparation with sweep cultivator for control of weeds.

— Sowing with seed drill preferably having fluted roller metering device for better plant stand establishment.

Recommended Crops and Varieties

Pearl millet - RHB 60, RHB 67, MH 179, WCC-75

Sorghum - Jowar-CSH-1, CSH-5, SPV-96

Grain legumes	- Mung bean	- S-8, S-9, K-851, PS-16
	Dew gram	-Jadia, T-2, T-18, Maru moth, Jwala
	Cowpea	- FS 68, K-11, Charodi-1, HFC-424 (for fodder)
	Cluster bean	-Durgapura safed, Maru guar, HG 75, FS 277
Oilseeds	-Sesame	-T-13, TC-25, RT 46
	Castor	- Aruna, Bagya, Gauch-1
	Mustard	- T-59 (Varuna), Pusa bold

Intercropping Systems

2 rows of mung bean + 1 row of pearl millet

2 rows of cluster bean + 1 row of pearl millet

2 rows of sesame + 1 row of cluster bean/dew gram/ mung bean

Soil Fertility Management

Rotation	-Kharif to kharif rotation of millets with legumes
Pearl millet	- 40 kg N ha^{-1} in two equal splits; 20 kg as basal and 20 kg as top dress
Pearl millet	- Legume rotation - 20 kg N ha^{-1} to pearl millet as top dressing, 40 kg P_2O_5 ha^{-1} to legume.
Oilseeds	- 30 to 40 kg N ha^{-1} coupled with 40 kg P_2O_5 ha^{-1} for sesame and sunflower.

Plant Protection

Romoval of weeds within 25 to 30 days of sowing either by cultural practices or by using weedicides is useful. Pre- emergence application of 0.5 kg a.i. of atrazin ha^{-1} in pearl millet and herbicides (alachlor, basaline and nitrofen) for grain legumes, are recommended. For the control of insects, pests and diseases, recommended pesticide schedule may be followed.

Contingency/Alternative Crop Planting

The principal source of water for dryland crops, north west monsoon, is highly unrealiable in terms of quantum, distribution, onset and withdrawal. In view of this following alternate cropping strategies are recommended to meet the unprecedented weather aberrations.

(a) *Delayed onset of monsoon (late in July)* : Transplanting 18 to 20 days old seedlings of pearl millet on limited areas. Diversion of major area earmarked for pearl millet to *kharif* pulses.

(b) *Delayed onset of monsoon (upto first week of August)* : Sowing of traditional grain legumes - cluster bean, dew gram and mung.

(c) *Occurrence of early drought* : Gap filling of pearl millet by transplanting to maintain adequate plant population. Complete removal of weeds for conservation of limited moisture in the soil profile to prolong the crop survival.

(d) *Occurrence of late drought* : Reducing of plant population within the rows and removing of alternate crop rows in the event of severe drought. Protective irrigation from harvested water either at critical growth stages or at any time before maturity to mitigate the adverse effect of the drought.

Although this technology can act as an insulation against the moderate moisture stress, occurrence of frequent droughts causes serious hindrances in its adoption at farmer's level. This is so because the risk of negative or insignificant returns under the abnormal situations persists. However, studies on cropping systems, combining perennial component with the annual crops, have revealed that the risk factor is to an acceptable level (Daulay and Bhati, 1989).

FARMING SYSTEMS FOR SUSTAINABILITY

An average family in the Indian arid zone comprises of 7 to 8 members with 6 to 8 hectares of land holding and 20 to 22 livestock heads, mainly cattle, sheep and goat. About 60 to 70 per cent of the farmers live in the village itself while their farm lands are located 2 to 3 km away. Only 30 to 40 per cent have their dwellings at the farm and thus have better control on their land and vegetation. Therefore, farming systems development (FSD) approach under these two situations is different i.e. arable farming system for the former category and integrated farming system for the later.

Arable Farming Systems

In arid areas, cropping system and the crops (even varieties) behave differently in different years because of erratic monsoon (C.V. 40 to 70 per cent). Therefore a farming system encompassing diversified crops and cropping systems as per the rainfall models vis-a-vis feed, fodder, fuel and cash requirement of the farm family, has to be chosen. Grown on 60 per cent of the land holding as a sole or mixed crop pearl millet is a staple food crop of this region, though vulnerable to drought and late onset of monsoon (beyond 15th July). Therefore, pearl millet to be taken in 40 per cent of the land holding and remaining 60 per cent area can be put under grain legumes (30 per cent), oilseeds (15 per cent) and fodder crops and forage grasses (15 per cent).

Among the perennial components, only grasses have some scope as they can withstand grazing pressure and play significant role in the soil conservation i.e. wind filter strip, bund stabilizer, vegetative barrier etc. *Seccharum bengalensis, Panicum antidotale, Panicum turgidum, Cenchrus ciliaris, Cenchrus setigerus* and *Lasiurus sindicus* are the potential grasses for this purpose. Diversification of crop components in the arable farming system is not only sustainable but also provide multitude products and by-products to meet requirements of food and fodder. Further, the system can sustain about 21 adult cattle unit (ACU) ha^{-1} with average body weight gain of 350 g day^{-1} $animal^{-1}$.

Integrated Farming Systems

For the farmers who can manage their lands intensively throughout the year, integration of alternate land use systems viz. agri- horticulture, agri-pasture, agroforestry, farm forestry and silvi-pasture in a suitable proportion, is advocated to attain the high yield plateau with the upgradation and conservation of various natural resources (land, water, vegetation and animal).

An economic evaluation of these integrated land use systems over arable farming was carried out taking 18 years as the effective period. All these land use systems showed higher benefit-cost ratio (1.46 to 1.87) than the dryland crop production (1.24). To study the model integrated farming system in its totality for soil plant animal interaction, the components of the farming systems were bifurcated as agroforestry (25 per cent), farm forestry (20 per cent), agri-horticulture (20 per cent), agri- pasture (15 per cent) and silvi-pasture (20 per cent) in 1990 as per average land holding (7 ha). Silvi-pastoral system gave substantially high net returns (Rs. 5097.44 ha^{-1}) over the arable farming system (Rs. 2197.00 ha^{-1}) even in the second year of establishment. Besides, it could sustain 554 animal days ha^{-1} $year^{-1}$ (Tharparkar breed) with 60 per cent pasture utilisation.

Integrated Farming Systems in Watershed : A Case study

Under the national policy of development of drylands on

watershed basis, Jhanwar model watershed project near Jodhpur was undertaken jointly by CAZRI, Jodhpur and Department of Soil Conservation, Rajasthan in March 1987. The total area covered under the watershed was 4600 hectares scattered in five villages; Bujawar (38 per cent), Jhanwar (35 per cent), Chokha (21 per cent), Rohilla Kalan (4.7 per cent) and Doli (1.2 per cent). The upper reaches of the watershed covered 38 per cent of the area which is rocky, degraded and highly undulating (15 to 25 per cent slope). Several streams originate from the ridge area, causing severe to moderate erosion hazard to the adjoining arable farm lands, resulting in heavy degeneration of resources. A sizeable area of the watershed lies in class III (1102 ha), IV (1233 ha) and VI (273 ha) and requires adequate land treatment measures. The soils of the area are sandy in texture with very poor fertility. Although the mean annual rainfall of the region is 360 mm, it was sub-normal between 1987 and 1990, and therefore, perennial components in the integrated land use systems suffered a set back. The farmers of the area have a poor economic base with average land holding of 3.3 ha to support an average family size of 8.29 members. Based on the assessement of the resource and socio-economic conditions of the area a master plan was prepared. The project envisaged various hardware, software and alternate land use systems alongwith programmes of human and animal resource development.

Programme Implementation

In the first phase (1987-92), 1200 ha area of the watershed was undertaken for development. In the degraded and rocky catchment areas, three masonry dams were constructed to reduce the flow rate of water in the 'Nala' to check the erosion losses in the adjoining cultivated fields. For stabilization of gullies, several brushwood and dry stone check dams were erected. In the whole catchment area, silvi-pastoral system with *A.tortilis, A.senegal, Z.rotundifolia* alongwith *C.ciliaris* was taken up and protected by stone wall fencing. Encouraging results were also obtained in the community grazing land in the watershed area where reseeding of improved grasses increased the forage productivity by 5 to 6 times. Ditch-cum bund fencing, alongwith key line plantation of 3 row of *A.tortilis* was also found satisfactory for protection from the biotic interferences. Farmers showed keen interest in the dry farming technology developed at CAZRI particularly in harvesting and recycling of rainwater techniques for use in agri-horticulture, improved crops and cropping pattern and farm forestry. The component wise progress of the project over the five years (1987 to 1992) is given in Table 18.

TABLE 18. Progress of the Jhanwar watershed management project (1987 to 1992)

Technology	Total target (ha)	Achievement upto March 1993 (physical)	Cost incurred upto March 1992 (Rs in lakh)
Farm forestry	550	185 ha	0.42
Reclamation of gullied land by silvi-pastoral system	200	120 ha	1.50
Pasture development	150	60 ha	4.04
Improved crops and distribution of inputs on subsidy	25 ha, 5 ha/yr	25 ha	1.15
Masonry dams	3 nos.	3 nos.	1.30
Loose rock check dams	100 nos.	135 nos.	0.16
Brush wood check dams	80 nos.	70 nos.	0.12
Stabilizing 'nala' bands by tree plantation	10000 m	6000 m	0.40
Farm ponds	12 nos.	12 nos.	2.56
Animal health cover	-	-	0.17
Solar energy	-	42 solar cooker	0.25
Agri-horticulture	60	30	1.57
Total			14.04

CONCLUSIONS

Development of drylands on watershed basis, of which integrated farming systems are the indispensable components, is beyond doubt the only answer for the sustained productivity and restoration of ecological balance in the arid areas. However, there are socio- economic, technological, institutional constraints in the effective implementation of such programmes and there is an urgent need to overcome these so as to get the whole hearted participation of the people in execution as well as subsequent maintenance of the sustainable land use systems.

REFERENCES

Bhati, T.K. 1993. Integrated land use systems for drylands. Regional Training course on Land Resources and Their Management for Sustainability in Arid Regions. UNESCO-ICAR, CAZRI, Jodhpur, pp. 1-14.

Daulay, H.S. and T.K. Bhati. 1989. Crop and pasture production potential. *In* Review of Research on Sandy Soils in India, Intern. Symp. on Managing Sandy Soils, pp. 245- 261.

Daulay, H.S., T.K. Bhati and P.R. Ojasvi. 1990. Development and management of watersheds in arid areas - A case study. Proc. III IWRS National Symposium on Watershed Development and Management, pp. 284-294.

7

Shelterbelt Management for Environmental Improvement in Hot Arid Ecosystem

A.K. Sharma and J.P. Gupta

The major constraints to crop production in hot arid regions of India are low and erratic rainfall, extreme temperatures, high and desiccating winds, high evaporative demand of the atmosphere and the movement of sand. In the hot weather period (March to June), the wind movement increases and averages about 20 to 25 kmh^{-1}, although, maximum wind forces are several times greater thar the average wihd speeds. High wind causes (i) excessive evapotranspiration, (ii) uprooting and lodging of crops, (iii) soil erosion and (iv) heat injury by hot winds. These conditions lead to low level of crop production and in extreme cases (towards Jaisalmer and Bikaner) even crop failure. Therefore, a technology which is able to reduce the wind speed can be useful in improving the early growth and establishment of plants in this region.

DESIGN AND ESTABLISHMENT OF SHELTERBELTS

A shelterbelt is defined as a belt of trees and/or shrubs maintained for the purpose of shelter from wind, sun, snowdrift etc. Shelterbelts are generally more extensive than wind breaks and cover larger area than a single farm, sometimes a whole region on a planned pattern. The effectiveness of different shelterbelts, however, depends on its age, type, conopy growth, design and geometry.

When considering shelterbelt planting, three zones can be recognized; the windward zone from which the wind blows; the leeward zone (the side where the wind passes); and the protected zone in which the effect of shelterbelts is felt.

The effectiveness of the shelterbelt is determined by its permeability. If it is dense like a solid wall, the air flow will pass over the top of it and cause turbulence on the leeward side due to the lower pressure on that side, this gives comparatively a limited zone of effective shelter on the leewards side compared to the zone that a moderately permeable shelter creates. Optimum permeability is 40 to 50 per cent of open space, corresponding to a density of 50 to 60 per cent in vegetation crops in the barriers should be avoided. Permeability of dense shelterbelt can be improved by pruning lower branches at 1.50 to 0.8 m from the ground level.

It is generally accepted that a shelterbelt protects an area over a distance upto its own height on the windward side and upto 20 times its height on the leeward side, depending on the strength. When designing a shelterbelt the direction of the wind must be considered. A shelterbelt should be established perpendicular to the direction of the prevailing wind for maximum effect. For protection of large area, number of separate barriers can be created as parts of the whole system. When prevailing winds are mainly in one direction, a series of parallel shelterbelts perpendicular to that direction should be established, a check board pattern is required when the winds originate from different directions. The shape of the cross-section of a shelterbelt determines the sheltering effect to a large extent. The choice of tree or shrub species to plant, alongwith their planting arrangement dictates the cross-sectional shape. In general, an inclined shape facing the wind should be avoided as it only deflects the upward wind flow. Barriers with a clear vertical side provide best reduction in wind spreed. In reducing wind speeds, narrow barriers can be as effective as wide ones. Furthermore, a narrow shelterbelt has the advantage of occupying less land. Before establishing shelter belts, it is important to make a thorough study of the local winds and to plot a map of the directions and the strength of the winds.

Selection of Species

For selection of tree and shrub species for wind breaks or shelterbelts, following characteristics should be sought :

(1) Rapid growth

(2) Straight stems

(3) Firmness against wind

(4) Good crown formation

(5) Deep root system which does not spread into nearby fields

(6) Resistant to drought

(7) Desired phenological characteristics (leaves throughout the year or only part of the year) (Anon., 1989).

Planting

Planting techniques for shelterbelts are identical to those for other trees or shrubs. However, as shelterbelt plantations require a high survival rate as well as uniform and rapid growth, supplemental irrigation may be needed during the establishment phase. Gaps, if any, must be replaced promptly.

Although in theory, one row barriers should suffice, but experiences have shown shelterbelts consisting of several row of trees are most effective. Initial spacing required is 3 m between row to row and 2 m between the trees. When trees or shurbs have horizontal spread of roots to interfere with crops in the agricultural fields, veritical root pruning either by digging trenches or with some special equipment is recommended. A triangular arrangement of planting is recommended for most conditions.

In early sixties, shelterbelts of *Acacia nilotica* and *Dalbergia sissoo* planted at Central State Farm, Suratgarh (Rajasthan) were found very effective in controlling the sand movement.

Three-row wind breaks with *Acacia tortilis, Cassia siamea* and *Prosopis juliflora* as side rows and *Albizzia lebbek* as central row established at the Central Arid Zone Research Institute, Jodhpur were found effective. The details of observations on wind speed, evaporative demand, and soil water and nutrient status of soil taken after 8 years of plantation and 5 metres tall shelterbelts are discussed under the head "effects of shelterbelts". The trees take a lot of gestation time in establishment and growth till they form an effective shelterbelts and therefore, have a long term importance, "Micro-shelterbelts", on the other hand, have immediate and short term importance, it includes tall fast growing plants or crops which can be raised well in advance of the cropping season in irrigated condition. However, under rainfed conditions fast growing bushes or grasses can be raised in the first year and the short growing crops in the following years. Micro-shelterbelts of pearl millet used for cowpea and okra crops have been found effective (Gupta et al., 1983).

Management

Once established, the effectiveness and longivity of a shelterbelt depend on its maintenance. As the trees and shrubs grow, they change in shape and appearance, which necessitates some level of maintenance to ensure continued shelter effect. Pruning helps to stimulate height, growth, while thinning boosts diameter growth. To keep a barrier at the desired density and permeability, occasional pruning or removal of plants may be necessary. In all of these causes, the management practices depend on the desired composition of the barrier and the species used. Since these management practices can involve the removal of woody parts, the use of tree or shrub species that make fuelwood or fodder available on a continuous basis is desirable.

The life of shelterbelt is dependent on the trees or shrubs of which it is composed. Therefore, for permanent shelter, a renewal plan should be adopted. To renew a barrier consisting of many rows, felling the rows on the leeward side and then replanting them is often recommended. If the shelterbelt consist of one row, first a new row should be planted parallel to the old one; and when it attains the required growth the old one is removed. When shelterbelts are established on grasslands or other areas where animals are allowed to graze, special attention must be paid to the protection of the barrier. This can be done either by planting thorny vegetation or by using a barbedwire fence along the edges of the barrier.

EFFECTS OF SHELTERBELTS

Favourable Effects

Reduction in wind velocity and wind erosion : The results of the experiments conducted at CAZRI on the basis of daily wind speeds measurement showed that the maximum wind speed reduction at 2H distance (H is the average height of shelterbelts) and effectiveness upto 10H. There was 50 per cent reduction in wind erosion as compared to the unprotected field during summer (Gupta and Sharma, 1993).

Conservation of soil and water : In arid areas the evaporation is more influenced by aerodynamic term than the energy balance term, potential evapo-transpiration (ET) values calculated using Penman's methods showed a decrease of 8 to 20 per cent in ET rate in the leeward side at a distance of 2H and at 5H and 10H the decrease was 6 to 10 per cent and 2 to 6 per cent respectively. Shelterbelts have been found to reduce evaporative loss of moisture from the soil surface by reducing wind speed and providing favourable micro-climatic conditions. Gupta et al. (1983) observed that in 0 to 15 cm and 15 to 30 cm soil layers there was generally higher moisture content in sheltered area than in the unsheltered area (Table 19). *Cassia siamea* shelterbelt was, however, found to be more effective in conserving moisture and thus in delaying the drying of the soil. This could be due to more leafy structure of this shelterbelt which reduced the air permeability and conserved the moisture in soil. The total nutrient loss was also found to be the maximum from unsheltered soil followed by shelterbelts.

TABLE 19. Effect of shelterbelts on soil moisture, nutrient status and erosion

Type of shelterbelt	Nutrient loss (g ha^{-1})			Per cent soil loss of the bare soil	Per cent moisture in 0-30 depth 5 days after irrigation (%) of the bare soil
	Nitrogen	Phosphorus	Potassium		
Prosopis juliflora	123.0	65.0	64.1	63.6	112.1
Cassia siamea	64.5	34.1	350.1	33.6	126.3
Acacia tortilis	113.0	45.5	570.0	54.6	85.7
Bare soil	191.0	101.2	1039.0	100.0	100.0

Environmental Improvement

Several climatic factors, such as temperature, precipitation, humidity, wind, evapo-transpiration etc., are influenced to some extent by shelterbelts.

- Reduction in the daily range of variation in air temperature
- Increase in the humidity in the atmosphere
- Trees due to their crowns, interrupt the rain and cusion its impact on the ground and its erosive power. They reducé surface runoff and increase infiltration which help in building a large underground reservoir.
- Shelterbelts increase dust precipitation due to reduction in wind velocity and increase in relative humidity. One hectare of shelterbelt can filter 50 to 100 tonnes of dust.
- All above favourable effects i.e. reduction in temperature, water losses and wind velocity, increase the crop yield in the effective area of shelterbelts. Experiment conducted by ICRISAT in Africa shown that pearl millet yield increase by 20 per cent in crops grown between young (5 years old) shelterbelts, spaced about 100 m apart (Vanden Beldt, 1989).

· Shelterbelts can necessarily be a good source to yield timber, firewood, fodder and fruits. A single row shelterbelt planted along the field bunds yielded about 2.5 m^3 ha^{-1} yr^{-1} wood (Anonymous, 1989).

Adverse Effects

Resource competition : Shelterbelts may compete with the crop for growth resources viz. water, light and nutrient. This can be minimised by partial pollarding of shelterbelts or by making trenches along the shelterbelts to reduce competition for light and water.

Allelopathic effect : Some tree species are used for shelterbelts release some phytotoxins into the soil which adversely affect the germination and yield of the companion crops. In most cases, the allelopathic effects are relative and vary with trees and crops. In general, leaves are the most potent source of allelochems. However, the toxic metabolites are also distributed in all other plant parts in different concentrations. In some species of *Acacia*, tenin is found which has toxic effect on the nearby crop. There was 66%, 25% and 5 per cent reductions in the moong yield at 5 m, 10 m and 15 m away from *A. tortilis* shelterbelts (Saxena and Sharma, 1996).

REFERENCES

Anonymous 1989. Arid Zone Forestry, FAO Conservation Guide No.20, FAO, Rome.

Gupta, J.P., G.G.S.N. Rao, B.V.R. Rao and G.N. Gupta. 1983. Soil drying and wind erosion as affected by different types of shelterbelts planted in the desert region of western Rajasthan. J. Arid Environ. 96 : 58-63.

Gupta, J.P. and A.K. Sharma. 1993. Role of shelterbelts/wind breaks in arid zone agriculture. Proc. Nat. Sem. on Agroforestry for 2000 A.D. Jhansi, India, 94-96.

Saxena, A. and A.K. Sharma. 1996. Allelopathic potentials of *Acacia tortilis* in agroforestry systems of arid regions. Allelopathy J. 3(1): 81-84.

Vanden Beldt, R.J. 1989. Agroforestry in semi arid tropics. Strategies and Research Needs. ICRISAT Conf. Paper No. CP 399, Patancheru, Hyderabad, India.

8

Management of Degraded Grasslands in Arid Region of Rajasthan

K.C. Singh

Crop production in the hot desert of Rajasthan is unstable due to unfavourable climatic conditions. Pasture development for livestock raising has a very important role to play in the economy of this region. But due to heavy pressure of livestock on pasture lands the productivity has considerably decreased. The fodder requirement of livestock in this region is always in short supply due to increased livestock population(Venkateswarlu et al., 1992). Therefore, use of improved techniques for the rehabilitation of degraded grasslands is essential for achieving higher production. The present paper describes the results of different aspects of grassland management viz. methods of reseeding, fertilizer use, grass-legume-mixtures, silvi-pastures, agri-pastures and irrigation to grasslands in "Indira Gandhi Nahar Pariyojna" area.

RESEEDING

Sowing the seed mixtures of *Cenchrus* sp. + *Lasiurus sindicus* and *Cenchrus* sp. + *Dichanthium annulatum* has been found advantageous. For large scale development of pasture in arid regions, establishment of *Lasiurus sindicus* and *Dichanthium annulatum* was better with transplanting of rooted slips or seedlings as compared to direct seeding. However, where transplanting is not practicable on large scale due to water scarcity, pelleting of seed is most appropriate method. Pellets of 0.5 cm diameter are prepared by mixing grass seed with cow dung, clay and sand in a ratio of 1:1:3:1 using sufficient quantity of water. For direct sowing, the seeds of *C. ciliaris* should be mixed with moist soil in 1:3 or 1:4 ratio drilled uniformly in 8 to 10 cm deep and 50 cm apart furrows and covered with less than 2 cm soil layer. Such sown pasture needs 1 to 2 weedings during first year of its establishment (Singh, 1988, 1992; Shankarnarayan and Singh, 1990).

FERTILIZER USE IN PASTURE GRASSES

Application of fertilizer to pasture grasses during normal rainfall years increased forage as well as seed yield. Application of 20 kg N ha^{-1} to *Cenchrus ciliaris* (358) pasture resulted in 83.1 and 112.6 per cent increase in the dry forage and seed yields, respectively while phosphorus had no effect on the yield (Rao and Singh, 1994; Singh, 1988, 1992).

The protein content of fertilized pasture was also higher than that of unfertilized. Under irrigated conditions, application of 20 kg N ha^{-1} to grasslands of perennial grasses boost forage production. Harvesting of grass at flowering stage is necessary to get more production for better nutritive forage. Planting of tree species like *Acacia tortilis* on the boundaries of grasslands as wind break is also important. *Prosopis cineraria* and *Ziziphus* sp. are good top feed trees for plantation in such grasslands. Grazing should not be allowed in irrigated grasslands, rather these grasslands should be kept either as reserve or harvest grass for fodder banks to meet the requirement during drought years or lean periods (Singh, 1991).

GRASS LEGUME-MIXTURE

Introduction of *Stylosanthes hamata* after light working of soil in the established pasture of *Dichanthium annulatum* could increase the productivity of grassland in rainfall zone above 400 mm. The productivity was highest in *D.annulatum* + *S.hamata* system (Shankarnarayan and Singh, 1990) (Table 20).

TABLE 20. The effect of grass-legume mixture on forage yield and protein content

Grass species	Yield (kg ha^{-1})		Crude protein (%)	
	Pure	Mixed	Pure	Mixed
Stylosanthes hamata	1410	860	12.53	11.42
Dichanthium annulatum	3310	3330	3.81	4.18

AGRI-PASTURE

Studies were conducted at CAZRI, Jodhpur on agri-pastoral i.e. cropping between grass strips laid out at right angles to the prevailing wind direction. The average yields of *Vigna radiata* (mung bean) and *Vigna aconitifolia* (moth bean) obtained under unprotected plots were 272 and 257 kg ha^{-1}, which increased to 484 and 372 kg ha^{-1}, respectively, when the crops were protected by perennial strips of *Lasiurus sindicus* established at right angles to the wind direction. In addition to grain, 1650 kg ha^{-1} dry forage yield of *L.sindicus* was obtained in the protected plots (Mishra, 1964). Similarly, the average seed yield of cluster bean under protected plot was 418 kg ha^{-1} which increased to 503 kg ha^{-1} when the crop was protected by perennial grass strips. During low rainfall years, cluster bean crop failed and could not produce seed while grass produced 726 kg ha^{-1} dry forage. Even in severe drought year (1987), cluster bean failed but grass component produced forage. The average yield of *Cenchrus ciliaris* and *Lasiurus sindicus* were 2 and 4 t ha^{-1}, respectively. The amount of soil blown during summer months from unprotected plots was higher than those protected by grass strips (Singh, 1989).

Pearl millet was grown in the plots which were under grass and cluster bean consecutively for 7 years (1981 to 1987). The mean grain yield of pearl millet in grass plot was 848 kg ha^{-1} as compared to 565 kg ha^{-1} obtained in the cluster bean plot. Organic carbon content of crop plots in 0 to 15 and 15 to 30 cm soil layer was 0.16 and 0.14 per cent, respectively where as it was 0.19 and 0.17 per cent respectively under grass plots. Grass plots also contained more soil moisture than that of crop plots (Singh, 1995).

MOISTURE USE EFFICIENCY OF PROMISING GRASSES

Rao et al. (1993) reported that *Cenchrus* sp. create favourable micro climate with in canopy by maintaining lower temperature (2 to 9°C) and high humidity (3 to 11 per cent). The total consumptive use of *Cenchrus* sp. varied between 164 and 363 mm. The energy use efficiency of *C.ciliaris* was higher than *C.setigerus* during low rainfall year where as with increase in rainfall the energy use efficiency of *C.setigerus* was higher than that of *C.ciliaris*.

The consumptive use of *L.sindicus* varied between 144 to 271 mm with a mean value of 201 mm with water use efficiency ranging from 13.24 to 20.85 kg DM ha^{-1} mm^{-1}. *L.sindicus* utilized thermal energy efficiently under adequate moisture condition as indicated by the heat use efficiency values which varied from 0.60 to 0.77 kg

DM ha^{-1} $^{o}C^{-1}$. It was also found that *L.sindicus* utilized the thermal energy efficiently for higher production of forage than *Cenchrus* species.

CHOICE OF TREE SPECIES FOR ARID GRASSLAND

Prosopis cineraria is a known drought hardy top feed tree which grows well on pastureland without any detrimental effect on associated grass. The tree is lopped during winter season for its protein rich (about 14 per cent) leaves. *Ziziphus nummularia* also grows well on cultivated fields and grasslands and provides nutritious leaf forage (14 to 16 per cent protein). *Acacia tortilis* is suitable for wastelands development and sand dune fixation. *Calligonum polygonoides* is a sand dune fixation plant and its tender stem is useful top feed for animals. *Acacia senegal* grows well on rocky sites and provides leaves and pods for sheep and goats. *Prosopis juliflora* performs well on all types of soils and good for live fencing of pasture lands. It also provides nutritious pods. *Acacia nilotica* grows well on medium to heavy textured soils and provides nutritious leaves and pods. Other top feed species viz. *Capparis aphylla, Colophospermum mopane, Dichrostachys nutan* and *Hardwickia binata* also perform well on arid grasslands (Singh, 1988, 1992, 1995).

Forage yield of grass under *Prosopis cineraria* trees in sown pasture of *Cenchrus ciliaris* reduced significantly upto 1 m distance from the tree trunk. The dry forage yield of *C.ciliaris* obtained upto 1 m distance from the tree trunk was 1160 kg ha^{-1}. But the yields at 2 to 3 m distance was 2880 kg ha^{-1} as compared to 3280 kg ha^{-1} from open area at 15 m distance from *Prosopis cineraria* which were statistically at par. In general, grass seed yield increased with the increase in distance from the tree (Jain et al., 1989).

Forage production under 10 years old plantation of *Acacia tortilis, Dichrostachys nutan, Colophospermum mopane* and *Azadirachta indica* was studied at Borunda (Jodhpur). *Cenchrus setigerus* was sown in interspaces after establishment of trees. There was no significant difference between the dry forage yield of *Cenchrus setigerus* grown in association with *C. mopane* (1843 kg ha^{-1}), *D. nutan* (2073 kg ha^{-1}) and that in pure grassland (2287 kg ha^{-1}). But the yields in association with *A. tortilis* (839 kg ha^{-1}) and *A.indica* (1496 kg ha^{-1}) were significantly lower than those of other tree species and pure grassland. *A.tortilis* showed 97.6 per cent survival followed by *C.mopane* (90.5 per cent), *A.indica* (39.3 per cent) and *D.nutan* (40.5 per cent). Maximum height was recorded in *A.indica* (425 cm) followed by *A.tortilis* (393 cm), *C. mopane* (301 cm) and *D. nutan* (261 cm) (Singh, 1995).

The yield of *Cenchrus ciliaris* sown in between the rows of 9 fodder tree species planted in 1984 at a spacing of 10 m row to row and 4 m plant to plant was studied at Bhopalgarh (Jodhpur). Survival percentage, height and collar, diameter of trees recorded after 7 years showed maximum survival of *A.tortilis* and *D.nutan* (98 per cent) followed by *A.senegal* (88 per cent), *C.mopane* (82 per cent). *A.indica* (75 per cent) and *Z.nummularia* (65 per cent). *A.tortilis* showed maximum height (258.5 cm) followed by *A.indica* (188.8 cm) and *A.senegal* (143.7 cm). *A.tortilis* also showed maximum collar diameter (6.1 cm). The mean forage yield of grass under different tree species did not differ significantly and it was 2556 kg ha^{-1} as compared to pure pasture (3216 q ha^{-1}) (Singh, 1988, 1995).

SUPPLEMENTAL IRRIGATION TO *L. sindicus* AND *C. ciliaris* GRASSLANDS

L. sindicus is most adaptable grass species in low rainfall areas of Jaisalmer, Barmer and Bikaner districts of Rajasthan. It grows well on sandy plains and sand dunes. However, drought and low plant population are the important reasons for its low production. Hence the first step to increase the forage production is to bring the plant population to an adequate level through some management techniques. Increasing levels of sprinkler irrigation have been found to increase the yield of established pasture of *L.sindicus*. The water use efficiency was 37 kg DM ha^{-1} mm^{-1} with (92 mm of supplemental irrigation).

The water use and production potential of sprinkler irrigated *L.sindicus* pasture was studied on a sandy loam soil of Jodhpur using six irrigation levels i.e. 100, 90, 80, 55, 32 and 0 per cent replacement of water deficit in combination with 0, 40, 80 and 120 kg N ha^{-1} (Singh et al., 1990). Rooted slips of grass were planted in August 1985 at a spacing of 100 cm between row to row and 50 cm between plants. The soils of the experimental site had 114 mm available soil moisture in 1 m profile. First cutting of grass was taken in November and subsequently 4 cuts were taken at monthly intervals from March to June. In the plot with 100 per cent water deficit replacement, there was 93 per cent depletion of available water before cutting in November and 11 to 4 mm depletion in available soil moisture during summer and, no wilting symptoms were observed in grass even at such high moisture tensions. The treatment receiving 80 kg N ha^{-1} in combination with 90 per cent deficit replacement of available water (about 100 mm water) yielded

25.1 t ha^{-1} of green forage and 8.8 t ha^{-1} dry forage. Water use efficiency was related with soil fertility. It declined with the application of more water under low fertility, and increased steadily up to the use of 400 mm water, followed by an increase at a constant rate of 42 kg green fodder mm^{-1} of irrigation upto 700 mm, which was the total irrigation requirement to maintain the irrigated grass land of *L.sindicus* throughout the year under high fertility (80 kg N ha^{-1}) (Singh et al., 1990).

The response of *Lasiurus sindicus* and *Cenchrus ciliaris* to irrigation in different months was studied on a sandy loam soil of Jodhpur. Grasses were established by sowing of seed at a spacing of 75 x 50 cm in August 1986. Irrigation was given after each cutting from March to June. Grasses were harvested at about one month after irrigation at flowering stage and a total of six cuttings were taken. The total rainfall received from June 1987 to March 1988 was 77.3 mm.

The green forage yield of *L. sindicus* was maximum in August (5.4 t ha^{-1}) followed by February (4.3 t ha^{-1}) cutting. Dry forage yield was also highest (3.3 t ha^{-1}) in August with one supplemental irrigation. The total green and dry forage yields of six cuttings obtained from June to March were 19.0 and 8.7 t ha^{-1}, respectively. *C. ciliaris* gave highest green forage yield in February (7.4 t ha^{-1}) followed by August (5.3 t ha^{-1}). The total green forage yield from six cuttings was 23.7 t ha^{-1}. Dry forage production was also highest (3.4 t ha^{-1}) in February followed by August (3.0 t ha^{-1}) cutting. Total dry forage production from six cuttings was 11.0 t ha^{-1} (Singh, 1991). *C. ciliaris* can produce more green and nutritive forage throughout the year with less irrigation (Kavia and Singh, 1989).

These results indicate the irrigation schedule i.e. the first cut of rainfed grass should be taken at the end of September. Sprinkler irrigation of 100 mm may be given in October and November and grass should be cut at flowering stage at the end of each month. Third 100 mm of irrigation may be given in mid February and cutting can be taken by middle of March. The fourth 100 mm of irrigation is to be given in March and grass should be cut in the end of June. This will provide protection to the land surface from wind erosion since the wind velocity remains very high during April-June. The water requirement of *L. sindicus* for the whole year (4 cuttings) is about 400 mm. With this amount of water, *L. sindicus* can produce 10 t ha^{-1} dry forage per year. The carrying capacity of irrigated grassland on year-long basis would be 3 adult cattle units per hectare as against 0.5 adult cattle unit under rainfed grasslands.

OTHER FODDER CROPS

Apart from native grasses, Napier bajra hybrid also holds promise for growing on good soils with irrigation. An experiment on fertilizer use (50 and 100 kg N ha^{-1}) and intercropping of other fodder crops in Napier bajra Hybrid (NB-21) was conducted at Jodhpur in 1978-79. Cowpea, chinese cabbage and *Subabool* grown in the interspaces of Napier bajra hybrid gave an additional dry forage yield of 7.1 and 0.8 t ha^{-1} in 1978 and 2.9 and 1.5 t ha^{-1} in 1979 without significantly affecting the yield of principal crop. However, intercropping of lucerne with Napier bajra hybrid significantly reduced the yield of principal crop in both the years. The yield of Napier bajra hybrid with 100 kg N ha^{-1} increased significantly over the control. The maximum yield of Napier bajra hybrid under irrigation was 15.2 t ha^{-1} in association with cowpea and chinese cabbage (Singh et al., 1984).

CONCLUSIONS

The forage production of desert grasses can be increased significantly by adopting improved technologies of grassland management. In those areas of the desert where intensive irrigation is neither safe nor feasible, the limited water of Indira Gandhi Canal command area can be utilized by sprinkler irrigation to grassland of *L. sindicus* adjoining to canal command. Irrigation level should be 100 mm per month to be given in October, November, February and March. The supplemental irrigation to the grassland of *C. ciliaris* can also be given to get green fodder with less water throughout the year particularly during drought years. Apart from yielding palatable and nutritive forage grasses will also help in conserving the vulnerable soil resource in the fragile arid ecosystem.

REFERENCES

Jain, B.L., K.C. Singh, S.S. Rathore and R.N. Prasad. 1989. Influence of tree canopy on characteristics and productivity of loamy sand soils in arid region. International Symposium on Managing Sandy Soils, pp. 598-601.

Kavia, Z.D. and K.C. Singh. 1989. More forage of *Cenchrus ciliaris* with less water. *KHETI*.43 (3) : 21-22.

Mishra, D.K. 1964. Agronomic investigations in arid zone. Proc. Symposium on Problems of Indian Arid Zone, pp. 165- 169.

Rao, A.S., K.C. Singh, Y.S. Ramakrishna and R.S. Singh. 1993. Micro-climatic impacts on the relative growth of *Cenchurs ciliaris* and *Cenchrus setigerus*. Ann. Arid Zone 32 : 245-250.

Rao, A.S. and K.C. Singh. 1994. Influence of meteorological factors on forage and seed productivity of *Cenchrus ciliaris*. Ann. Arid Zone 33 : 39-44.

Shankarnarayan, K.A. and K.C. Singh. 1990. Rehabilitation of degraded pasture land in the arid region of Rajasthan. In : Technologies for Wasteland Development (I.P. Abrol and V.V. Dhruvanarayana, Eds.), ICAR Pub., pp. 21-31.

Singh, Mahendra, R.S. Paroda and T.K. Bhati. 1984. Studies on inter-cropping in napier bajra hybrid. Ann. Arid Zone 23: 347-352.

Singh, K.C. 1988. Management of arid rangelands of Rajasthan. In: Wastelands Development and Their Utilization (K.A. Shankarnarayan, Ed.), Scientific Publishers, Jodhpur, pp. 287-293.

Singh, K.C. 1988. Rehabilitation of degraded pasture land in the arid region of Rajasthan. In: Rangelands Resource and Management (Panjab Singh and P.S. Pathak, Eds.), pp. 72-76.

Singh, K.C. 1989. Importance of grasses in the management of sandy soils. Strip cropping of grasses. Intern. Symp. on Managing sandy Soils, pp. 561-564.

Singh, K.C. 1991. Production of forage grasses with minimal irrigation in western Rajasthan. In: Prospects of Indira Gandhi Canal Project (I.P. Abrol and J.Venkateswarlu, Eds.) ICAR Pub., pp. 82-86.

Singh, K.C. 1992. Grassland and pasture development in the arid region of Rajasthan. In: Rehabilitation of Degraded Arid Ecosystem (A.S.Kolarkar, D.C. Joshi and K.D. Sharma, Eds.), Scientific Publishers, Jodhpur, pp. 121-126.

Singh, K.C. 1995. Management of pasture lands in the arid regions. In: Sustainable Development of Dryland Agriculture in India (R.P. Singh, Ed.), Scientific Publishers, pp.415-423.

Singh, S.D., Y.V. Singh and K.C. Singh. 1990. Water use and production potential of sprinkler irrigated sewan (Lasiurus sindicus Henr.) pastures in Thar desert. Indian J. Agric. Sci. 60: 23-28.

Venkateswarlu, J., K. Anantha Ram, M.L. Purohit, M.S. Khan, H.C. Bohra, K.C. Singh and B.K. Mathur. 1992. Forage 2000 AD - The Scenario for arid Rajasthan. CAZRI, Jodhpur, pp. 1- 32.

9

Nutrient Cycling in Agroforestry

B.M. Sharma

Nutrient cycling processes, though take place to varying degrees in all land use systems, have special significance in the agroforestry context because of the role of trees on such processes. Nutrient cycling refers to the flow of nutrients into, through and out of an ecosystem. It involves geochemical, bio- geochemical and biochemical processes. Sum total of nutrients present in the system; their cycling time, turn over rates and resident time within the system; and biotic and abiotic factors which drive the nutrients are important aspects for understanding the nutrient cycling.

NUTRIENT POOLS

Soil

Soil is a three phase system consisting of solid, liquid and gases. The solid phase is reservoir of most of the plant nutrients in which clay and humus are the most reactive components. Weathering of parent material and organic matter release nutrients to soil pool which also works as sink for N- fixation and applied fertilizers (Nair, 1984; Young, 1989). Under arid conditions, high temperatures favour rapid mineralisation of organic substances and therefore organic matter status of arid soils is very low. The mean organic matter content in Rajasthan soils below 300 mm rainfall zone range between 0.05 to 0.2 per cent in sandy soils, 0.2 to 0.3 per cent in medium textured soils and 0.3 to 0.4 per cent in fine textured soils (Dhir, 1977). High temperature and low rainfall conditions also accelerate the loss of nitrogen and limit its accumulation in soil. Deforestation, soil erosion and intensive land use coupled with low level of fertilizer addition virtually result in poor soil pool. Total phosphorus in the light textured soils of arid zone have been reported to be 285 ppm as compared to 327 to 450 ppm in soils of semi arid regions (Ram Deo and Ruhal, 1972). These soils, however, are generally well supplied with K. Water soluble and exchangeable K together form 1.5 to 2 per cent of total K.

Vegetation

Vegetation pool has considerable importance in nutrient cycling, particularly with respect to soils of low fertility (poor soil pool). Perennial vegetation (trees and shrubs) has greater significance in nutrient cycling as compared to annual crops. In perennial vegetation, nutrients remain locked up for longer time (Naraian and Singh, 1985; Rawat and Singh, 1989), there is annual addition to the vegetation pool by way of growth (Bornmisza, 1982) and tender parts and leaves of trees contain more amount of nutrients. Deciduous trees translocate nutrients from leaves to perennial organs before leaf fall. On the other hand, due to frequent harvest of annual crops the amount of nutrients which enters the system is much higher than the removal as a result practically there is no recycling. Nutrient content in foliage of some important tree species of arid zone is reported in Table 21.

TABLE 21. Nutrient content (%) in foliage of important tree species of arid zone

Tree species	Nutrient				
	Ca	Mg	K	Na	P
Prosopis cineraria	1.85	0.31	1.40	0.058	0.169
Acacia nilotica	1.96	0.22	1.11	0.066	0.178
A. senegal	4.80	0.69	1.33	0.065	0.153
A. tortilis	2.16	0.35	1.44	0.058	0.137
Salvadora oleoides	2.67	0.54	1.10	0.377	0.091
Tecomella undulata	2.24	0.49	1.28	0.078	0.169
Ailanthes excelsa	3.16	0.75	2.05	0.068	0.414
Prosopis juliflora	3.18	0.53	1.51	0.085	0.208
Albizzia lebbek	2.67	0.40	1.23	0.035	0.147
Azadirachta indica	3.05	0.50	1.49	0.056	0.194

Source : Dhir et al. (1984)

Litter

In perennial vegetation and grassland ecosystems, litter and dead roots are important nutrient store (Turner, 1988). The leaf litter or the pruned material when incorporated into the soil release nutrients on decomposition. The total litter produced in a unit time, its decomposition rate and loss (erosion) determine the amount of litter present and the nutrient status of soil (Murthy et al., 1990). The decomposition rate of leaves, however, varies with tree species (Sharma, 1993). In a tropical environment, the ratio of litter production to accumulation is wider. Addition of leaf litter through different tree species of arid zone is given in Table 22.

TABLE 22. Leaf litter addition through different trees

Tree species	Litter g m^{-2}
Prosopis cineraria	69.0
Tecomella undulata	39.2
Albizzia lebbek	23.2
Acacia senegal	30.3
Prosopis juliflora	23.0

Source : Venkateswarlu (1989)

Although the litter store is small, it is renewed annually with some variations depending on the climate. Leaf litter is related more closely with N-cycling as nitrogen is bound up in organic molecules in the soil humus. In arid zone, a major portion of the above ground biomass of vegetation is consumed and addition/accumulation of litter and its contribution to nutrient cycling is meagre.

NUTRIENT FLUXES

Movement and relocation of nutrients with in the soil plant system is a biogeochemical process. The main flows of nutrients are from plant component to plant residues via soil fauna to soil humus, through the process of mineralisation, and return to the plants via root uptake. Efficient management of nutrient cycling is related with uptake, return and reduction of nutrient losses. The amount and rate of nutrient uptake determine the nutrient flux (Fig. 2).

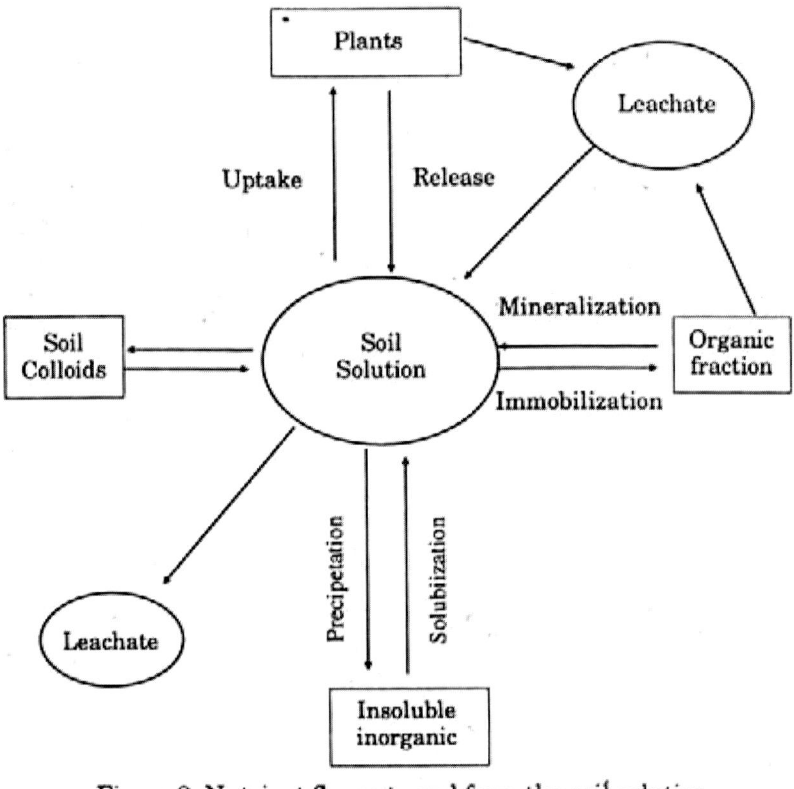

Figure 2. Nutrient fluxes to and from the soil solution

INPUTS AND OUTPUTS TO THE SYSTEM

INPUTS

Nitrogen Fixation

Nitrogen fixation takes place through non-symbiotic and symbiotic means. Non-symbiotic N-fixation or azotification is carried out by bacteria, blue green algae and some fungi. These organisms derive energy from decomposing organic matter and therefore dependent on the organic matter status of soil. The symbiotic N-fixation occurs through bacteria living in close association with the host plant (legume) and fix nitrogen in the root nodules. Good aeration, optimum moisture, good supply of Ca and Mo and low initial N levels are conditions favourable for these organisms. Nitrogen fixation by some tree species is given in Table 23.

TABLE 23. Nitrogen fixation by trees and shrubs

Tree species	N-fixation kg ha^{-1} yr^{-1}
Acacia albida	20
Casurina equistifolia	60-110
Glicirdia sepium	13
Leucaena leucocephala (6 months)	100-130
Prosopis glandulosa	25-30
Prosopis tamarugo	200

Source : Nair (1984)

These values are approximate, as there are some inherent problems in methods of measurement i.e. nitrogen by difference, acetylene reduction and N^{15} labelling. Inclusion of leguminous trees in the agroforestry system not only fix nitrogen into the soil but also helps to reduce root competition with crops. Leguminous trees can meet their N requirement partly through fixation and thus more nitrogen becomes available to the interplanted non-nitrogen fixing crops.

Weathering of Rocks

Weathering of parent material release nutrients to become available to plants and therefore a principal source of nutrients. A nutrient if not present in the parent material no level of cycling can add it except from external source. Deep root system of trees enhance the uptake of nutrients by weathering and reduce their fixation by clay minerals.

Precipitation

Nutrients can also be added to the system through precipitation (Nelson et al., 1982). Suspended dust in the atmosphere is principal source of mineral nitrogen in the rain water. On an average, approximately 5 kg N ha^{-1} yr^{-1} enters the soil through precipitation. Under arid conditions of Rajasthan, 1.84 to 6.33 kg ha^{-1} ammonical and 3.1 to 5.89 kg ha^{-1} nitrate nitrogen has been found to be added through precipitation (Aggarwal et al., 1982). Vegetation cover and the amount of rainfall influence the nutrients added to soil (Switzer et al., 1988). In *Acacia albida*-groundnut system, out of 3 kg ha^{-1} N received with 1200 mm rainfall 2 kg was received as throughfall and 1 kg was retained by tree canopy (Araguven et al., 1982).

Manures and Fertilizers

Organic manures such as farmyard manure and compost are the mainstay in fertility management. However, the supply of organic manures in arid regions is too limited to make any significant contribution. Organic manures may be added indirectly by the cattle dung in a silvipasture system. Application of small amount of organic manures in the planting pits of trees helps for initial establishment and better growth add nutrients to the soil. The crops grown in the system fertilized as per the recommended practices add directly to the nutrient pool as left over after meeting the seasonal demand of the crop. However, in arid zone due to scanty and erratic distribution of rainfall generally very little quantity of fertilizer is applied and therfore addition to the nutrient pool is also low.

Pruning

This is most important flux involving the return of nutrients to soil pool which can be managed as input. But in arid zone, most of the pruned material is used as fuel, fodder, food etc. and incorporation to soil is negligible except small twigs and leaf as left over in the field.

OUTPUTS

Almost all the nutrients are liable to losses but nitrogen is more mobile and removed from the system easily. The processes of nutrient loss include gaseous, leaching, erosion etc.

Gaseous

Volatilization of ammonia is one of the most important avenues of nitrogen loss in arid zone. Temperature, moisture, pH, clay and $CaCO_3$ influence the volatilization losses of nitrogen. Alkaline pH accelerate the process and losses can be as high as 22 per cent of the applied nitrogen (Sharma et al., 1991). In coarse textured soils of arid regions where the atmospheric temperatures are high nitrogen losses upto 20 per cent may occur. However, application of urea as band placement and gypsum to alkali soil can substantially decrease the losses.

Burning of plant residue is another important way of gaseous losses of nutrients. Most of carbon, nitrogen and sulphur held in the plant residue are lost in this way. Although phosphorus, potassium and calcium are retained in the ash and can subsequently be released into the soil, loss of these nutrients is likely to occur in particular form, ash being carried up by heat and blown away by wind.

Leaching

The water percolating through the soil take away the nutrients present in the soil solution beyond the root zone and thus make them unavailable to the growing plants. Soil texture, amount of water available for leaching and several other factors determine the extent of losses through this process. In sandy soils, an increase in the depth of water increase the urea movement (Praveen Kumar et al., 1990) which was closely related with rate of urea nitrification (Praveen Kumar et al., 1989). The losses of phosphorus from the system may not be high because of its less mobility in the soil. Under arid conditions, because of limited availability of water, the losses through the process of leaching is rather low.

Soil Loss

Removal of soil particles through the action of wind/water carry attached nutrients from the fields. Organic matter, being the lightest constituent in soil, is easily picked up and removed. Removal of nutrients through moving particles is different and varies with soil texture, vegetation cover and management practices. Loss of N occurs mainly through organic matter and very little in the ionic forms. While P is lost through clay and organic matter both. Large amounts of K is removed through erosion but loss of Ca and Mg are insignificant. In arid regions, nutrient losses are mainly through wind erosion and severe. Removal of 1 kg surface soil could result in a loss of 0.35 g N, 0.185 g P and 1.90 g K. Vegetation cover decreases the nutrient removal through this process.

Harvest of Produce

Accumulation of nutrients in biomass of trees is more than the annual crops. Therefore, harvest of whole tree is a major drain of nutrients from the system. Frequent pruning of trees for fuel purpose removes considerable amount of nutrients from the system. Since the harvest of annual crops is quite frequent, removal of nutrients is significant. That is why continuous arable agriculture with low inputs decline the fertility of soil. The amount of nutrient removed per unit quantity of the produce harvested varies greatly with the species due to inherant differences in their capability of nutrient uptake. For example, nitrogen and phosphorus uptake by moong is higher than cluster bean. Apart form the individual species effect, tree-crop combinations and their proportionate arrangements also affects the nutrient removal. The uptake of nutrients by two grasses grown under the tree canopies presented in Table 24 shows that the removal of nutrients through grasses in association with trees was lower than the grass alone. A similar decrease in uptake of pearl millet and cluster bean crops grown in association with trees is also observed. Nitrogen uptake by safflower, cabbage, oat and barley crops grown under the canopy of 4 to 5 year old *Albizzia lebbek* decreased by 44, 40, 44.7 and 29.7 per cent respectively over the sole crops (Hazra and Singh, 1986; Tripathi and Hazra, 1986). Thus the trees not only retain the nutrients by themselves but also helps to reduce the nutrient removal through crop harvest by decreasing the uptake.

TABLE 24. Nutrient uptake by two grasses grown in association with trees

Tree grass combination	Kg ha^{-1}				
	P	K	Ca	Mg	K
C. ciliaris + *A. tortilis*	2.0	10.2	8.9	3.6	1.5
C. ciliaris + *Z. rotundifolia*	2.3	11.8	10.3	4.2	1.7
C. ciliaris alone	2.7	13.6	11.8	4.8	2.0
L. sindicus + *A. tortilis*	2.6	15.3	27.0	7.7	0.6
L. sindicus + *Z. rotundifolia*	3.0	17.7	31.2	8.8	0.7
L. sindicus alone	3.4	20.0	35.2	10.0	0.8

CONCLUSIONS

Agroforestry systems lead to more closed nutrient cycling by improving the ratio of inputs and outputs. The objectives in designing and managing agroforestry systems should be to make more efficient use of nutrients. It is desirable to reduce the ratio between inputs/outputs such that the nutrients are reused by plants before being lost from the system.

REFERENCES

Aggarwal, R.K., P. Kaul and A.N. Lahiri. 1982. Ammonia and nitrate in rainwater in the semi arid areas of western Rajasthan. Ann. Arid Zone 21 : 299-305.

Araguven, J., G. Escalente and R. Herrera. 1982. Nitrogen cycles of tropical perennial crops under shade trees. Plant Soil 67 : 247-269.

Bornemisza, E. 1982. Nitrogen cycling in coffee plantation. In : Nitrogen Cycling in Ecosystems of Latin America and the Carribean (G.B. Robertson, R. Herrera and T. Rosswall, Eds.), The Hague, pp. 241-46.

Dhir, R.P. 1977. Western Rajasthan soils : Their characteristics and properties. In : Desertification and Its Control, ICAR, New Delhi.

Hazra, C.R. and S.B. Tripathi. 1986. Influence of N on some soil properties, and forage production of safflower and chinese cabbage with and without tree association. J. Indian Soc. Soil Sci. 34 : 275-280.

Murthy, I.Y.L.N., C.R. Hazra and Ashok Kumar. 1990. Effect of incorporation of tree leaves on soil fertility. J. Indian Soc. Soil Sci. 38 : 325-327.

Nair, P.K.R. 1984. Soil productivity aspects of agroforestry. ICRAF, Nairobi.

Narain, Pratap and Ravindra Singh. 1985. Behaviour of nitrates under different land uses. Indian For. 111 : 230-39.

Nelson, L.E., G.L. Switzer and M.G. Shelton. 1982. Bulk deposition and leaching losses via base flow from sandy upland soils, upper coastal plains, Mississippi. Mississippi Acad. Sci. Biloxi, USA.

Praveen Kumar, R.K. Aggarwal and B.M. Sharma. 1988. Nitrification and nitrate movement from ammonium sulphate and urea in sandy soils as affected by their previous application. Fert. Res. 21 : 29-36.

Praveen Kumar, R.K. Aggarwal and B.M. Sharma. 1990. Effect of time and depth of water on movement of urea and its distribution as ammonium in an arid sandy soil at two levels of initial soil moisture. J. Indian Soc. Soil Sci. 38 : 394-403.

Ram Deo and D.S. Ruhal. 1972. Status and availability of phosphorus in calcareous soils of Rajasthan. Indian J. agric. Sci. 42 : 316-321.

Rawat, Y.S. and J.S. Singh. 1989. Forest floor biomass, litterfall and nutrient return in central Himalayan Oak forests. Vegetation 82 : 113-125.

Sharma, B.M., R.K. Aggarwal and Praveen Kumar. 1991. Ammonia volatilization as influenced by amendments in a light textured alkali soil. Ann. Arid Zone 30 : 203-207.

Sharma, B.M. 1993. Annual Report. CAZRI, Jodhpur.

Sharma, B.M. 1994. Annual Report. CAZRI, Jodhpur.

Switzer, G.L., L.E. Nelson and M.G. Shelton. 1988. Influence of canopy of Lobolly pine plantation on the deposition and chemistry of precipitation. Mississippi State Univ. USA Tech. Bull. No. 154.

Turner, J. and M.J. Lambert. 1988. Impact of wild life grazing on nitrogen turnover in the understorey of a Eucalyptus forest Aust. For. 51 : 222-225.

Venkateswarlu, J. 1989. Strategies for harnessing natural resources in arid west Rajasthan. Indian Fmg. 39 : 19-20, and 31.

Young, A. 1989. Agroforestry for soil conservation. CAB Intern. and ICRAF, Oxon, U.K.

10

A Conceptual Model for the Evaluation of Agroforestry Systems

P.R. Ojasvi and R.K. Goyal

The basic requirements of an agroforestry system are that its components should be complementary to each other, the system should be sustainable vis-a-vis soil conditions and it should be economic. Several hypothetical procedures based on the human experiences are available in literature to fulfil these conditions. However, examples of analytical or experimental support are very few. Agroforestry research is complex and takes long time to come out with reliable experimental results. Therefore, several authors have proposed modelling studies in order to understand agroforestry system behaviour.

This paper describes a conceptual model for analysis of agroforestry system. The concept of the model is drawn from discussions made in several papers about resource sharing theory, problems therein, system's potential for amelioration of soil condition and its projection as an economically viable system. A system methodology was applied to agroforestry in order to develop an analytical model. The structure of the full model and that of its sub-systems is presented and discussed. The model has been formulated with the objective to understand and analyse the overall system behaviour. In order to achieve this objective, the model will concentrate on the following aspects :

(1) To explore suitable and compatible tree-crop combinations for agroforestry system design

(2) To make predictions about the soil conditions under a specified system

(3) To test the economics of a compatible and sustainable agroforestry system, and finally to realize the concepts of the model

(4) To suggest the immediate research needs in agroforestry.

STRUCTURE OF THE MODEL

An abstract representation of the agroforestry system model is presented in Fig.3 . The purpose of the model is to give 'good' or 'best' tree-crop combination that can be implemented in a real situation. The agroforestry system can basically be defined by three components i.e. TREE, CROP and SOIL, CROP contains both food crops and grasses. Although, by definition agroforestry system also includes ANIMAL as a component but as our focus is to determine suitable tree-crop combinations, this model will not address the interaction of animals. Besides, it is difficult to mathematically model such interactions.

The model consists of three modules - Tree-crop interaction module (TCIM), Soil conservation module (SCM) and Economic evaluation module (EEM). Each module is a complete model in itself. The first module models the characteristics of TREE and CROP components of the system and constraints imposed by their interactions. Second module contains the SOIL component of the system and will model the changes in soil due to addition of litter, nitrogen fixation and erosion control. At this point, organic matter addition due to the presence of animals can be brought into the system. Separation of TREE/CROP components from SOIL is necessary owing to the differences in their modelling characteristics. The TCIM is based on the dynamic budgeting of the plant growth resources (nutrients, water and light), whereas, SCM can be based on the models of nitrogen, carbon or phosphorus cycles plus soil erosion prediction models. However, since it is a dynamic model, information generated in SCM will flow to TCIM and vice versa for their literative operation.

The starting point of the model is feedback from a diagnostic survey of the problem area in which type of agroforestry system is decided. The user of the model will then select a reasonable tree-crop combination from various available alternatives of tree and crop species for further analysis of the system. A reasonable selection of tree and crop species should be made in the very beginning (Huxley, 1985).

The interconnections of modules are imposed upon by an unique constraint. For example, interconnection between TCIM and SCM is constrained by compatibility, while SCM and EEM is constrained by sustainability. In other words, a performance criterion is

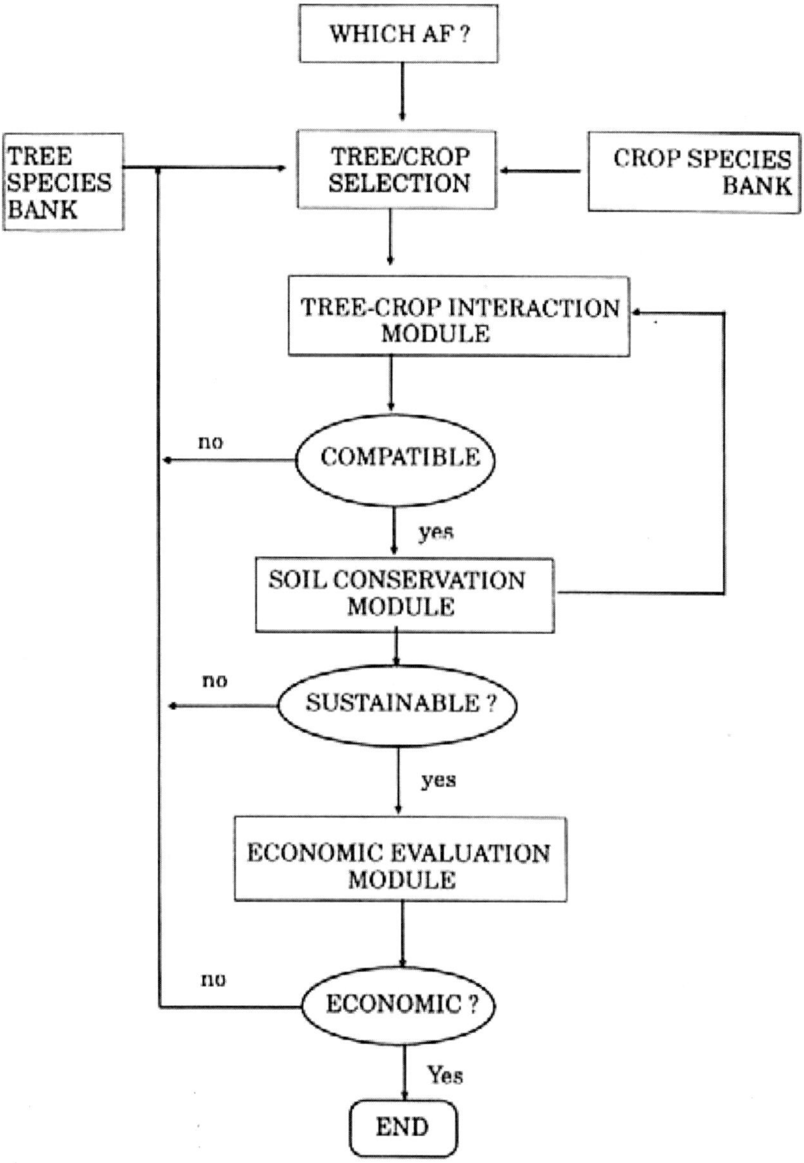

Figure 3 Time based agroforestry system model

available for the user to evaluate the system performance at every stage. The hypothesis here is that if tree-crop combination is not compatible initially, it will not ameliorate the soil condition in future, hence the model will invite the user to select another tree-crop combination without going into analysis for longer time periods. Similarly, if a particular tree-crop combination is not sustainable, then there is no use of its economic analysis.

COMPATIBILITY

In a field where trees are planted in association with crops, in the beginning (0 years), both TREE and CROP will deplete the resources from the same pool. As far as nutrients are concerned, it is only after the '*t*' Years, TREE component of the system will begin to show its benefits. In other words, from *o* to *t* years both TREE and CROP will function alike. Therefore, an obvious question arises that whether, up to *t* years and even after that (for light and water), TREE and CROP will grow without adversely affecting growth of each other or not. It is almost widely accepted that certain tree and crop species are complementary with respect to use of resources. Hence, compatibility can be defined as "the situation where optimum levels of resources are available to both the TREE and CROP components in their association". The optimum level here means physiological optimum.

However, this is not an adequate criterion for evaluating compatibility because it is possible that a tree crop combination initially incompatible for nutrients may become compatible after some time (as in case of N_2-fixing trees or fertilizer application). Also, certain plants have the potential to adapt to stress conditions, that cannot be accounted for by the fixed criterion like physiological optimum levels. However, this problem can be dealt with by setting optimum levels to an allowable 'low' level. This phase of the model will be interactive in which decision maker will make use of the model to develop management strategies and explore various alternatives.

SUSTAINABILITY

It can be defined as the "situation where land use systems retain their productive potential over a period of time, through conservation of the natural resources on which that productivity depends". Most important aspect of sustainability is the maintenance of soil fertility (Young et al., 1987). For modelling purpose, a simple but adequate criterion for sustainability is to check whether fertility level is in a steady state or not. Thus, if a tree-crop combination maintain soil fertility in a steady state over a long time period (may not initially), it is sustainable.

THE LOGIC

Tree-Crop Interaction Module : Light, water and nutrients are three major resources required for continuous plant growth. The optimum level of these resources is determined by their physiognomic growth habits and the environmental conditions. These resources can be thought of as pools, sharing of which by tree and crop components results in competitive, differential and complementary situations (Buck, 1986). Competitive situation may arise when one component has a preferential access to a resource, suppressing the need of other. The differential situation occurs when the resource availability to the plant components is divided in space and time. The complementary situation occurs when resource pool can fulfil the needs of all the agroforestry components at the same time and in the same space.

Considering the spatial configuration of tree and crop species as fixed, the requirement of these resources varies with time depending upon the growth responses of the two components. The TCIM will necessarily determine these two situation for a particular tree-crop association and not the differential situation. Before describing the model in detail, it is important to discuss the critical characteristics of these resources and their sharing criterion in agroforestry systems.

Light : Solar radiation (Rs) is unstorable but inexhaustible in nature, varies widely with geographical regions and seasonal fluctuations. Solar radiation is instantaneously available and if not used immediately in photosynthesis is lost forever. This constraint implies that, in agroforestry preferential access of one component can seriously damage the growth of the other. For instance, radiation intercepted by a large tree canopy reduces its availability to the crop growing underneath. However, presence of trees does not necessarily mean serious restriction in the photosynthetic activity of the understorey crop (Connor,1983a) . The light requirement of many crops in their sole culture has been quantified by measuring photosynthetically active radiation (PAR). In agro-forestry, Buck (1986) reports an experiment where Transmission of PAR from few agroforestry systems was found to be more efficient in absorption of PAR as compared to that from young maize. Since agroforestry experiments take long time to establish, PAR for trees can be measured by selecting trees of a particular species of different age and sizes. Then it can be expressed mathematically and used in the model directly. However, agroforestry experiments should continue to collect more reliable information for incorporation in the model for more accuracy.

Water : The storage of water in the soil is semi-permanent as losses occur due to evaporation and percolation depending on the climate and soil type. It is highly a time variant resources. Water uptake by a plant depends on its root profile and efficiency of extraction. However, water requirement on a particular plant species can be measured in discrete time units, i.e. at various growth stages. Water requirement of crops is well quantified and for trees considerable research is in progress.

Nutrients : The availability of nutrients, unlike water, can be predicted with reasonable accuracy. Nutrient uptake by a plant is also a function of various environmental and biological factors. Nutrient cycling through plant and soil is a complex process but has been successfully modelled in many cases.

THE MODEL

Dynamic modelling of a system can be done either by differential or difference equations (Manetsch and Park, 1977). A block diagram of TCIM is presented in Fig.4. It should be noted that block diagram of TCIM does not represent a resource cycling approach. It is essentially a time derived differential dynamic model, which determine the state of resource pools after a discrete time interval. Both differential and difference equations can be used to solve this model, later being simple. Each line in the block diagram represents a variable and each block can be represented by mathematical equations, desirably by linear state equations. Sigma block represent a summation block. The integral block determines the changes in the state of the resources.

The input variable (exogenous to the system) to the light sub-system is PAR or Rs. Tree canopy is first to intercept the radiation and the remaining transmitted radiation is available for the understorey crop. Tree and crop demand blocks (TR and CR) receive feedback from the tree growth process (TGP) and crop growth process (CGP) blocks about the change in the requirement. A delay component is introduced in the feedback loop as the growth process takes time to respond. The TGP and CGP blocks are not complex growth models but simply determine the changing plant requirements over the time and can be represented either by mathematical equations or by a information system (data base). The input variable in water sub-system is rainfall or irrigation (Ws). Water pool block determines the soil storage function. From this pool losses occur in the form of evaporation and percolation. Rest of the

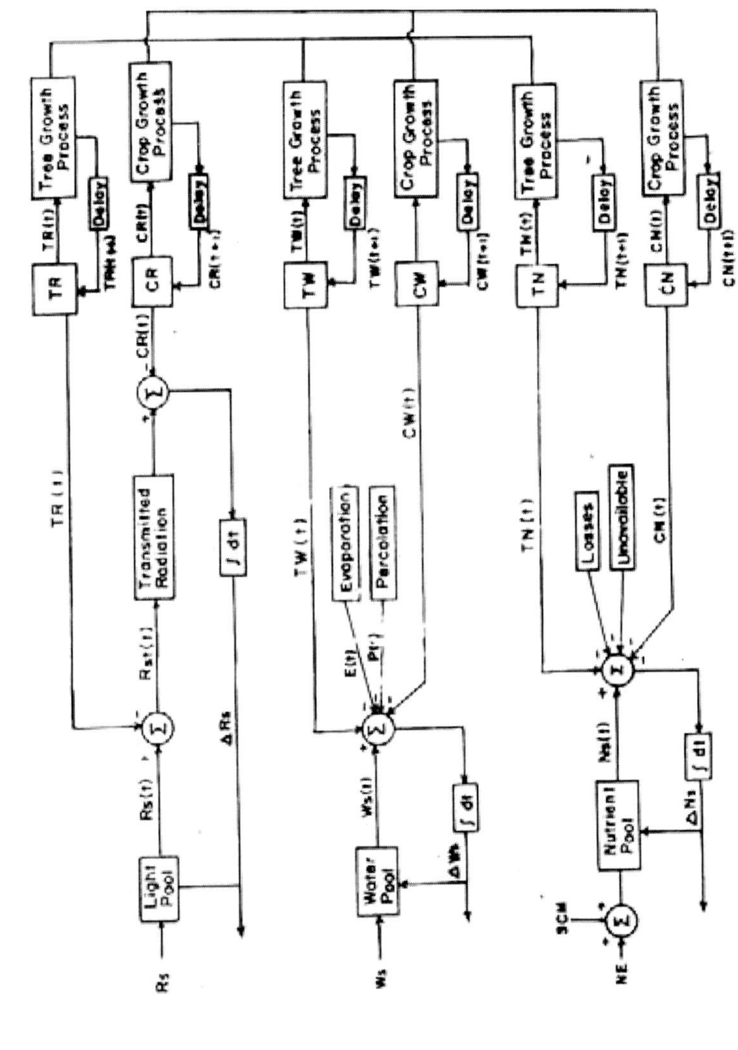

Figure 4. Block diagram of tree-crop interaction module.

block functions are as described above. The input variable to nutrient sub-system is initial fertility status of the soil. An additional information flows from the SCM which is improvement in the soil fertility due to litter decomposition and erosion control. The variables represented by the lines in the block diagram are state variables, i.e. that are internal to the system. The output variables are the final state of the resource pools and a compatible tree-crop combination with its spatial configuration.

So far, spatial arrangement of the tree and crop components was assumed to be constant and time was the only independent variable. In agroforestry associations, resource sharing varies with space also. Variations due to spatial configurations can be incorporated in the TGP and CGP which will supply information for the changing arrangements.

Drawbacks of TCIM : Firstly, it is based on the information available on the plants in monoculture and not in association while plants in mixed-culture behave differently than in sole culture. Resource sharing by plants in mixed-culture induces competition for resource which may change their growth behaviour. Response of plants to various stresses has been discussed in detail by Buck (1986) and Connor (1993 a). Secondly, the data available for the plants in sole culture are also inadequate.

SOIL CONSERVATION MODULE

Soil conservation in agroforestry encompasses both erosion control and fertility maintenance. Maintenance of soil fertility is one aspect where most agroforestry research activities are concentrated. Information is being generated for a wide variety of agroforestry systems (Fassbender et al., 1988). However, examples of a generalized computer model for predicting soil conditions under agroforestry systems are very few. A notable work has been done by Young et al. (1987). In this section, their SCUAF (Soil Changes Under Agroforestry) model is discussed to show that an approach is available for modelling soil conservation and can be capitalized for agroforestry system analysis.

SCUAF model covers the prediction of soil erosion and changes in soil fertility together with feedback effects of soil changes on plant growth. The present version of the model is operative for prediction of changes in organic matter and is being extended to include the cycling of specific nutrients, like nitrogen. The model is basically a dynamic carbon cycling model. It is based on the balance of carbon through the atmosphere, soil and plant compartments. The cycle is divided into plant and soil sections. The plant compartment consists of TREE and CROP components, that are divided into four parts, leaf, fruit, wood and root, where carbon is stored. Another carbon storage is the soil compartment, whole soil profile or the top soil only. The carbon sources are atmosphere and organic additions through litter decomposition and farmyard manure. The carbon sinks are atmosphere, erosion and various harvest of plant components. The model operates on an annual basis.

The SCUAF model assumes presence of established trees in the system. The outputs of the model are internal flows of the carbon within the model and its changes with time in soil and plant systems. The model is essentially concerned with the prediction of soil fertility but, as proposed by the authors, it can be used for designing agroforestry systems just only to determine whether a particular system will be sustainable or not. The SCM in the proposed system model can be replica of the SCUAF with slight modifications in its functioning, structure remaining the same.

ECONOMIC EVALUATION MODULE

Once a compatible and sustainable tree-crop combination is available, its economic analysis can be done with the usual procedures. The economic criterion is based on the benefit-cost ratio which should be equal to or greater than one. The equality criterion is applicable in the situations where invisible and unaccountable benefits can be more. A detailed discussion is not required here since economic procedures are well understood and require computerization only. This module is essential for the completeness of the system analysis.

CONCLUSIONS

A system model is presented which can be utilized for designing and understanding the behaviour of agroforestry systems over a time scale. Since agroforestry research is time consuming and complex and its experimental methodology is still under developmental stage (Huxley, 1989), models can provide important clues on the system behaviour. Initially, a simple approach is adequate for modelling purpose, more accuracy can be incorporated when specific information on tree-crop interaction is available. Similar views have been expressed by Connor (1983 b). Further, the present model suggests new directions for research in agroforestry over the traditional one. Following are the wide aspects which need attention :

— Resource utilization and their variation with time should be quantified for potential agroforestry trees and crops in their sole culture

— The above quantification should be done for different environmental and ecological conditions

Two reasons can justify above research activities. Firstly, it will be much less time consuming and easy to perform. Secondly, traditional agroforestry research is functionally complex and susceptible to imbalanced results due to even small disturbances. In other words, it is more difficult to establish a cause and effect relationship for tree-crop interaction and related aspects. This model will help in the generation of specific hypothesis in addition to an insight on the overall system behaviour.

REFERENCES

Buck, M.G. 1986. Concept of resource sharing in agroforestry system. Agroforestry Systems 4 : 191-203.

Connor, D.J. 1983a. Plant stress factors and their influence on production of agroforestry associations. In : Plant Research and Agroforestry (P.A. Huxley, Ed.), ICRAF, Nairobi.

Connor, D.J. 1983b. Crop models : Component of and contributors to models of agroforestry plant associations. In : Plant Research and Agroforestry (P.A. Huxley, Ed.), ICRAF, Nairobi, P. 249-256.

Fassbender, H.W., L. Alpizar, J. Heuveldop, H. Folster and G. Enriquez. 1988. Modelling agroforestry system of cacao (*Theobroma cacao*) with laurel (*Cordia alliodora*) and poro (*Erythrina poeppigiana*) in Costa Rica, Part-III. Agroforestry Systems 6 : 49-62.

Huxley, P.A. 1985. The basis of selection, management and evaluation of multipurpose trees - An overview. In : Attributes of Trees as Crop plants (M.G.R. Cannel and J.E. Jackson, Eds.), Instt. of Terrestrial Ecology, NERC, UK.

Huxley, P.A., T. Darnhofer, A. Pinney, A. Elijah and D. Gatama. 1989. The tree/crop interface : a project designed to generate experimental methodology. Agroforestry Abstracts 2 (4).

Manetsch, T.J. and G.L. Park. 1977. System analysis and simulation with application to economics and social systems. Pt. 1, Dept. Elect. Engg. & Sys. Sci., Michigan State Univ., USA.

Young, A., R.J. Cheatle and P. Muraya. 1987. Soil changes under agroforestry (SCUAF) : A predictive model. Working paper No. 44, ICRAF, Nairobi.

11

Role of Legumes alongwith Biofertilizers in Agroforestry Systems for Higher Plant Productivity

A.V. Rao

The basic limitations imposed by the physical environment in which plant species compete are natural resources. Factors such as light, carbon-dioxide, water and nutrients determine the ultimate performance of a farming system and success of a particular method of production whether it be based on a single species or a combination of several will depend on how effectively these resources are shared between the component plants.

Unlike others, light is a resource that cannot be stored for its subsequent use, it is effectively lost if not captured by the green parts (chloroplasts) of the plant. The dry matter production by a crop is determined to some extent, by the amount of light intercepted by the foliage when water and nutrients are available in sufficient quantities. With the available water and nutrient resources, 95 per cent of the incident light can be intercepted by growing crops in between trees having different canopy structure and also the leaf architecture. This particular concept may be taken into consideration for selecting a combination of tree and annual crops in agroforestry to minimise the producton losses resulting from light reaching the soil surface.

For complete utilisation of the captured light, the plants must be adequately supplied with water and mineral nutrients. The demand for water is primarily determined by the climatic factors. As evapo-transpiration is very high under arid climate the demand for water by a particular plant is more than that of the same species under humid climate. With the depletion of soil water reserves, the interplant competition develops which depending largely on the distribution and efficiency of the different root systems. It is possible to imagine the combination of plant species where deep rooted plants exploit water from reserves while the shallow rooted ones get water from the upper layers. By skilful selection of such species, water from the whole soil column can be utilised for higher plant production.

Legumes have been recognized as important components of agroforestry system. Apart from the direct benefits through nitrogen fixation to production, nitrogen contributed to the soil can be used by the companion crop. Further if a legume is grown with a non-legume, it can improve the nitrogen economy of the non-legume both by contributing N to the soil for uptake by the non-legume (referred as N- transfer) or simply by removing less soil nitrogen and thus leaving more nitrogen available to non-legume. The plant production under this system is based on "Facilitative production principle" in which the crops may alter the environment of one another positively for higher production (Vandermeer, 1989). On the contrary, two different plant species use different components of the ecosystem or use the same component in different ways by not competing with each other. The biomass production and nitrogen accumulation of *Cenchrus ciliaris* grass growing alongwith *Leucaena diversifolia* was enhanced compared with monocultured grass (Rao and Giller, 1993). Further, labelled nitrogen (^{15}N) was detected in both leaves and roots of intercropped grass and the percentage of grass nitrogen derived from *Leucaena* varied from 28 to 31.6 per cent. Nitrogen transfer may be due to root exudations, sloughed off root material or decomposition of nodule tissue. Haystead et al. (1988) observed that vesicular arbuscular mycorrhizal hyphal strands further help in translocating nitrogen from legume to non-legume.

The risk involved in the application of fertilizers in arid zone farming due to low and erratic rainfall the use of biofertilizers has been stressed upon. Biofertilizers are the microorganisms used in the inoculation of seed/soil for supplying nutrients in available form to the plants. Most common among these are nitrogen fixers and vesicular arbuscular mycorrhizae.

BIOLOGICAL NITROGEN FIXATION

Nitrogen, an important plant nutrient available for tree growth in the soil with the exclusion of applied fertilizer, is derived either from the mineralization of soil organic matter or biological nitrogen fixation (BNF). Nitrogen fixation occurs either symbiotically through legume-*Rhizobium* and *Casuarina-Frankia* (nonlegume - Actinomycete) association or non-symbiotically in the rhizosphere of trees. Inoculation of a suitable nonsymbiotic N-fixing bacterium associated with the tree roots is a possibility but research so far had mixed success.

The most successful management of soil microorganisms has been symbiotic nitrogen fixing bacterium, *Rhizobium*. Initially the appropriate management strategy was inoculation with selected rhizobia suited for a particular legume species and soil. Studies conducted on a number of leguminous trees have shown that trees contribute nitrogen to the soil ranging from 60 to 250 kg N ha^{-1} depending on the tree species, climate and soil type (Pokhriyal and Mathani, 1989). Effective nodulation of seedlings and nitrogen fixation is a prerequisite for better establishment and growth after transplanting. Among legume trees, BNF has been extensively studied in *Leucaena leucocephala*. Nitrogen fixation and biomass production of *L. leucocephala* upon inoculation with efficient strains of *Rhizobium* were better compared to uninoculated seedlings in the nursery, and establishment of inoculated plants was better than normal (Sanginga et al., 1989).

There is an urgent need to assess the performance of other legume tree species in the presence or absence of introduced rhizobia both under nursery and out plantations. It was found that inoculation with different strains of *Rhizobium* enhanced the nodulation and nitrogen fixing ability of 6 month old *Albizzia lebbeck* besides improvement in plant height and biomass production (Table 25).

TABLE 25. Nodulation and biomass production of *A. lebbeck* as influenced by rhizobial inoculation

Strain of *Rhizobium*	Plant height (cm)	Dry matter yield ($Plant^{-1}$)	No. of nodules	N_2 - ase activity
Control	25.3	7.8	4	2.8
Al_7	36.3	11.0	19	11.2
Al_8	40.3	15.0	44	17.9
Al_{14}	36.0	10.2	16	11.5
Al_{21}	42.8	16.8	44	18.5
LSD (P = 0.05)	2.0	1.2	3.1	2.1

* μ moles C_2H_4 $plants^{-1}$ h^{-1}

The inoculated plants were more healthy and could withstand stresses during transportation and out plantations. Umali Garcia et al. (1968) reported the improvement of nursery stock of *A. falcataria* and *Acacia mangium* upon inoculation with efficient strains of *Rhizobium*.

VESICULAR-ARBUSCULAR MYCORRHIZAE

Arid soils contain about 720 kg ha^{-1} total phosphorus but the available phosphorus is very low (7 to 15 kg ha^{-1}). To make phosphorus available in optimum quantities, phosphorus fertilizers which are expensive, are to be applied. Alternately either of phosphorus solubilizing microorganisms or vesicular arbuscular mycorrhizal fungi may be employed.

It is suggested that the mycorrhizal habit is evolved as a survival mechanism from both partners of the association allowing each to survive in the existing environment of low soil fertility, drought, disease and temperature extremes. The role of mycorrhizae in improving the quality of planting stock in the nursery and to obtain better seedling survival and growth after planting has been widely recognised all over the world (Cordell et al., 1987). Vesicular-arbuscular mycorrhizae (VAM) otherwise referred as endomycorrhizae occur over a wide range of ecosystems from aquatic to desert environments (Kiran Bala et al., 1989). VA-mycorrhizae are formed by non-septate phycomycetes fungi belonging to the genera *Glomus, Gigaspora, Acaulospora, Sclerospora*. These fungi are obligate symbionts and have not been cultured on nutrient media.

Degradation of land is often accompanied by salinization following improper changes in water regime. VAM fungal infection could overcome some detrimental effects caused by concentration of salts (Allen and Cunningham, 1983). These studies further indicated a possibility of using salt tolerant VAM fungi for rehabilitating saline areas. Potential involvement of tree mycorrhizae in direct nutrient cycling was demonstrated by Herrera et al. (1980). VA-mycorrhizae improve host's resistance to drought (Harley, 1969) and protect feeder roots from diseases (Marx, 1972).

In an extensive survey, Tarafdar and Rao (1990) observed variations in the intensity of VAM infections among different plant species collected from the same locality as well as within the plant species from different locations. The infection rate, in general, is dependent on age of the plant. Most of the VAM fungi present in arid soils belong to *Glomus, Gigaspora,* and *Endogone*. The growth and establishment of *Prosopis juliflora* was better upon inoculation with different VAM fungi compared to non-mycorrhizal plants (Table 26).

TABLE 26. Influence of VAM fungi on production of *P. juliflora* (2 year old in the field) and biochemical parameters in the rhizosphere (Tarafdar and Rao, 1993)

Parameters	Control	*Glomus mosseae*	*G. fasciculatum*
Plant height (cm)	133.5	189.7***	207.0***
Dry matter yield (kg plant^{-1})	1.7	4.1***	5.0***
Number of viable VAM propagules (50 g^{-1} soil)	66.6	187.6***	299.5***
Per cent root infection	21.6	81.8***	84.9***
Microbial populations:			
Fungi (x 10^3 g^{-1})	12.5	9.5*	14.5
Bacteria (x 10^4 g^{-1})	56.8	113.0***	136.5***
Actinomycetes (x 10^4 g^{-1})	8.0	12.0**	21.0***
Nitrosomonas (x 10^4 g^{-1})	0.47	0.52*	0.50
Organic matter (%)	0.29	0.36*	0.38**

*P = 5%, **P = 1%, ***P = 0.1%

Nursery and field studies indicated that selected VAM fungi on nursery seedlings reduce culls and improve field survival and growth (Cordell et al., 1987). VAM fungal isolates may also help *Leucaena* and *Prosopis* species to mitigate the adverse effects of sodium chloride on growth and development (Dixon et al., 1993). Upon inoculation with different VAM-fungi dry matter production of different plant species was found to be enhanced with maximum enhancement in *Prosopis juliflora* (Table 27).

Glomus fasciculatum was found to be the most effective isolate. VAM fungi not only improve the growth of trees but also help in revegetation of disturbed or degraded lands. It was recognised that mycorrhizae play an important role for a successful implementation of revegetation efforts. In general, mycorrhizal infection enhances plant growth by increasing nutrient uptake through increased absorbing surface area, mobilizing sparingly available nutrients, improving drought tolerance of plant by extracting water, through excretion of chelating compounds or ecto-enzymes and by exerting growth promoting effects. Mycorrhizal infection may also protect the roots from soil pathogens thereby increasing the healthy root growth.

TABLE 27. Effect of VAM fungi on the growth of forest trees (5 months old)

Treatment	No. of VAM spores 100^{-1} g soil	Plant height (cm)	Dry matter yield (g plant^{-1})
	Prosopis juliflora		
Control	677.5	15.6	1.3
Glomus mosseae	1288.3 (90.2)	26.9 (72.4)	3.5 (169.2)
G. fasciculatum	2092.0 (208.8)	41.4 (165.4)	5.2 (300.0)
	Albizzia lebbeck		
Control	771.0	8.8	2.3
G. mosseae	1396.5 (81.1)	9.7 (10.2)	2.9 (26.1)
G. fasciculatum	1557.7 (102.0)	9.9 (12.5)	2.9 (26.1)

Figures in parentheses indicate % increase over control.

Utilization of VAM fungi markedly increases the success of rehabilitation of disturbed and degraded lands either in mesic or in moisture deficient zones. It has been demonstrated that the presence of VAM is critical for the regeneration of natural ecosystems in arid lands. Inoculation techniques for rehabilitation have been tested and are considered highly valuable but until improved techniques of inocula production are developed, the best alternate is proper management of site, preventing loss and maintenance of an appropriate plant cover where the plant roots serve as an energy source and medium for VAM proliferation. Besides several advantages, these fungi are able to scavenge a number of nutrients from the soil for the uptake by the plants (Fig.5).

Legumes require adequate phosphorus supply for statisfactory nodule formation and nitrogen fixation (Van Schreven, 1958). VA -mycorrhizal fungi are known to improve plant phosphorus nutrition particularly in low P soils (Tinker, 1980). Inoculation of legumes with VA mycorrhizal fungi and *Rhizobium* certainly exerts synergistic effects on growth of legumes and it is attributed to the balanced nutrition of the host by mycorrhizal inoculation which results in enhanced nitrogen fixation by *Rhizobium* (Crush, 1974). Umali-Garcia et al. (1988) observed that dual inoculation with VAM and *Rhizobium* improved nodulation, nitrogen fixation and

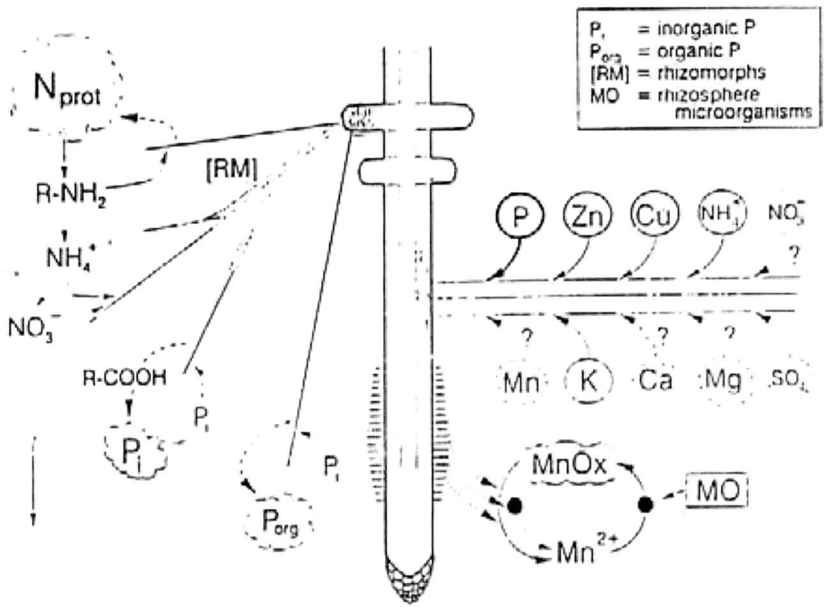

Figure 5. Schematic presentation of components of the nutrient dynamics in and acquisition from the "hyphosphere of endo-(AV)- mycorrhizal roots and of additional components found in ectomycorrhizal roots.

mycorrhizal colonization resulting in higher biomass production of *Leucaena, Acacia* and *Albizzia* in a P-deficient soil compared to single inoculation with either organism. Further, Dixon et al. (1993) reported that dual inoculation with VAM fungi and *Rhizobium* may help *Leucaena* and *Prosopis* species in mitigating the adverse effects of sodium chloride.

CONCLUSIONS

There is sufficient evidence in enhancing the productivity of tree seedlings upon inoculation with an efficient strain of *Rhizobium* or with any of the mycorrhizal fungi. Though there are opportunities to apply the microbial technology to seedling production and afforestation programmes both in degraded lands and reclamation sites, research efforts need to be directed towards finding the appropriate tree-microbe combinations adopted to well defined soil types and climatic conditions and the cultural techniques and suitable protocols in afforestation programmes on different sites. Trees and microorganisms are evolved together by adopting to natural and man made stresses. A major benefit of microbial associations is to help the host tree in managing these stresses. So more programmes aimed at selecting suitable organisms to improve stress management by tree are needed. The symbiotic relationship is an integral component of nursery seedling production. Regardless of the selected alternatives, nurserymen, field foresters and the planters must be aware that they are dealing with two symbiotic living organisms. Both must be nurtured to provide seedlings of superior quality for field plantations.

It may be necessary to broaden the education of our future agriculture researchers to cover a wide range of disciplines. The plant breeder in particular has an important role to play as almost all varieties of various crops have been developed in the context of sole cropping. New varieties must be developed/evaluated for agro-forestry or inter-cropping systems. The objectives of varieties selection or development are primarily to minimise competition and maximise complementary aspects. Canopy of the plant, leaf architecture, distribution and activity of roots etc. have to be kept in mind for developing varieties.

REFERENCES

Allen, E.B. and G.L. Cunningham. 1983. Effects of vesicular arbuscular mycorrhizae on *Distichlis spicata* under three salinity levels. New Phytol. 93 : 227-236.

Cordell, C.E., J.H. Owen and D.H. Marx. 1987. Meeting the challenge of the Nineties: Proceedings Intermountain Forest Nursery Association, August 10-14, Oklahoma city. Gen. Tech. Report RM-151, Fort Collins, Co.

Crush, J.R. 1974. Plant growth responses to vesicular-arbuscular mycorrhizae. VII. Growth and nodulation of some herbage legumes. New Phytol. 73 : 743-749.

Dixon, R.K., V.K. Garg, and M.V. Rao. 1993. Inoculation of *Leucaena* and *Prosopis* seedlings with *Glomus* and *Rhizobium* species in saline soil : Rhizosphere relations and seedling growth. Arid Soil Res. Rehabil. 7 : 133-144.

Harley, J.L. 1969. The Biology of Mycorrhiza, 2nd edition. Leonard Hill, London.

Haystead, A., N. Malajczuk and T.S. Grove. 1988. Underground transfer of nitrogen between pasture plants infected with vesicular-arbuscular mycorrhizal fungi. New Phytol. 108 : 417-423.

Herrera, R., C. Jordan, H. Klinge and E. Mediana. 1990. Amazon ecosystems : Their structure and functioning with particular emphasis on nutrients. Interciencia 3 : 223-232.

Kiran Bala, A.V. Rao and J.C. Tarafdar. 1989. Occurrence of VAM associations in different plant species of the Indian desert. Arid Soil Res. Rehabil. 3 : 391-396.

Marx, D.H. 1972. Ectomycorrhizae as biological deterrents to pathogenic root infections. Ann. Rev. Phytopath. 18 : 429-454.

Pokhriyal, T.C. and G.P. Mathani. 1989. Techniques and approaches to utilize nitrogen fixing system in forestry plantation programmes. Indian Forester, 115 : 136-158.

Rao, A.V. and K.E. Giller. 1993. Nitrogen fixation and its transfer from *Leucaena* to grass using ^{15}N. Forest Ecol. Manage. 61 : 221-227.

Sanginga, N., K. Mulongoy and A. Ayanaba. 1989. Effectivity of indigenous rhizobia for nodulation and early nitrogen fixation with *Leucaena leucocephala* in Nigerian soils. Soil Biol. Biochem. 21 : 231-235.

Tarafdar, J.C. and A.V. Rao. 1990. Survey of Indian arid zone tree species for the occurrence of VAM infections. In : Current Trends in Mycorrhizal Research (B.L. Jalali and H. Chand, Eds.), pp. 42-43.

Tarafdar, J.C. and A.V. Rao. 1993. Possibilities of manipulating mycorrhizal association for higher production in *P. juliflora*. In : Workshop on Potential of *Prosopis* spp. in arid and semi-arid regions of India, 21- 23 Nov., CAZRI, Jodhpur, India.

Tinker, P.B. 1980. Role of rhizosphere microorganisms in phosphorus uptake by plants. In : The Role of Phosphorus in Agriculture (F.E. Khasawneh, E.C. Sample and E.J. Kamrath, Eds.), p. 617-654. American Society of Agronomy, Medison, USA.

Umali-Garcia, M., J.S. Libuit and R.L. Baggayan. 1988. Effects of *Rhizobium* inoculation on growth and nodulation of *Albizzia falcataria* (L.) Fash. and *Acacia mangium* in the nursery. Plant and Soil 108 : 71-78.

Van-Schreven, D.A. 1958. Some factors affecting the uptake of nitrogen by legumes. In : Nutrition of the Legumes (H.G. Hallsworth, Ed.), p. 137-163, London.

Vandermeer, J. 1989. The Ecology of Intercropping. Cambridge University Press, Cambridge, p. 237.

12

Water Relations of Desert Vegetation

S. Kathju

The soil moisture conditions directly influence the utilization of water by plants. The close relationship of water use with plant production on one hand, and the hydrological cycle on the other, makes it an important area of management practice particularly in desertic area where precipitation is meagre and a close partitioning is required between water use by crops for grain, grasses for forage, trees for fuel and deep percolation for recharging the ground water. It is necessary to bear in mind that water use by vegetation vary widely depending on soil factors including soil moisture, climatic conditions and growth stage.

SOIL WATER

The sandy soils have low water holding capacity. All moisture present in the soil may not necessarily be available to plant. The field capacity (moisture retained in the soil after deep drainage/gravitational movement ceases after full saturation of the soil) is the upper limit of the moisture available for plant. Permanent wilting (moisture during soil drying when plants wilt unless water is added to the soil) is the lower limit of moisture in soil that is available for plant use. The wilting point is the water potential at which plant ceases to grow, but may recover once water supply is restored while ultimate wilting point is the plant water potential at which plants fail to recover when water is applied. Difference between these two is called wilting range (Furr and Reeve, 1945).

CELL WATER RELATIONS

The plant water status was expressed as the moisture content per unit weight of plant tissue. This was replaced by the relative water content (RWC) which is water content per unit of fully rehydrated tissue. Since 1960 moisture content in the soil and plant is described as energy content of water and is termed water potential. The water potential can be expressed as bars, atmospheres or dynes cm^{-2}. The most recent unit is Pascal (Pa/kPa/MPa). The water potential in plants and soils is negative, more so during drying phase. The plant water potential may be partitioned into three major components represented as under:

$$\psi w = \psi s + \psi p + \psi m$$

where :

ψw	=	water potential,
ψs	=	osmotic or solute potential - due to dissolved salts, addition of solutes to pure water reduces water potential, hence it is negative,
ψp	=	pressure or turgor potential - due to hydrostatic pressure on soil water or turgor pressure acting outward on cell wall, it has positive sign, and
ψm	=	matric potential - due to suction forces resulting from surface tension associated with the cell wall and colloidal surfaces which bind some of the water, it has negative sign.

The water potential of pure water is zero. It decreases (more negative) when solute is added. Thus water potential of cell sap is lower than that of pure water. When cell gains water the solutes are diluted which results in solute/osmotic potential becoming less negative. At this time wall pressure working against entry of water is created and thus turgor pressure increses. Under turgid condition, water potential is a function of osmotic and pressure potentials since matric potential approaches zero. Matric potential is small except in very dry tissues or dry seed. Under situations of soil moisture deficits, osmotic and matric potentials in plants increases and pressure potential approaches zero.

MOVEMENT OF WATER

Water moves from roots to leaves as a continuum through xylem. However, its movement from root epidermis to xylem and from xylem in leaves to cells and intercellular spaces is uncertain. It may move through: (i) vacuole, (ii) the symplast, (iii) walls or (iv) all three pathways. In slowly transpiring plants, the water may move through osmosis but in rapidly transpiring plants it is mainly due to tension or negative pressure in the xylem sap. Water is lost from leaf surface as vapour and is dependent on cuticular and stomatal resistance, and gradient in water vapour pressure between transpiring surface and outside atmosphere.

Water is taken up by plants both passively (in rapidly transpiring plants) due to transpiration and actively (in slowly transpiring plants). In arid and semi-arid plants where transpiration demands are high and plants lose moisture at a rapid rate contribution of active absorption is negligible. Apart from availability of moisture in soil, characteristics of soil and transpiration and rooting behaviour are the major factors regulating absorption of water.

WATER USE

Grasses : Due to shallow root system, growth of annual grasses and kharif crops is restricted to rainy season only which may extend from middle of July to September. Lahiri (1965) reported that after a rainfall of 163.8 mm the soil moisture in the profile upto 2 m depth beyond September was below the PWP and it was obvious that the ground cover consisting of *Dactyloctenium sindicum, Cenchrus setigerus, C. biflorus, Eleusine compressa, Crotalaria burhia, Aristida funiculata, Cyperus mutundus, Gisekia pharnaseolides, Brachiara ramosa* completely utilized the available moisture in the upper layers. Thus the non-availability of the soil moisture appears to be the main reason for cessation of growth beyond September. In the range, grasses like *Cenchrus ciliaris, C. setigerus* and *Lasiurus sindicus* the management conditions like soil fertility and herbage removal bear a close relationship with the water use. During a favourable rainfall year (1976 : 639.7 mm) the consumptive use in the pure stand of *C. ciliaris* (1 m x 1 m spacing) ranged from 152.2 mm to 162.2 mm. and in *L. sindicus* it ranged from 168.8 to 185.8 mm.

Trees : Most of the desertic trees transpire at a fast rate all round the year. In summer afternoons, the rate of transpiration may range from 1200 to 1500 mg g^{-1} hr^{-1} in many trees (Lahiri, 1980). Certain trees like *Salvadora oleoides* maintain a relatively low transpiration rate. In winter, water use is reduced mainly due to shorter day length and low temperatures. According to an estimate 222 mm of water was lost annually from one hectare of land having about 50 trees (Lahiri and Kumar, 1967). The desertic trees possess deep root system which enables them to tap moisture from deep moist soil layers. This often makes such trees independent of the moisture regime in the upper strata of the soil. But the situation in tree seedlings is different. The seedlings, depend on the soil moisture available in the upper layers due to restricted root system in the initial stages. Thus tree seedlings, unless irrigated, often experience soil moisture deficits (Burman et al., 1991). Under such situations seedlings maintain high transpiration per unit of leaf area, but economize through reduced surface area (Lahiri and Kharbanda, 1966). The leaf area is curtailed due to reduced expansion of leaves when stress is experienced during leaf expansion or enhanced abscission of the expanded leaves.

Crops : The crop water use is dependent on the atmospheric demands and the moisture availability in the root zone.The water use in legumes like clusterbean, mung bean and moth bean is low in comparison to pearl millet. The sum of consumptive use and other losses of moisture for the whole growing period may often found to be higher than the actual precipitation. Since an upward movement of water due to capillary is low in these sandy soils, the lateral movement of moisture or a vapour phase movement through temperature gradient may be possible cause of this phenomenon.

CROP RESPONSES TO SOIL MOISTURE DEFICITS

When transpiration exceeds the water absorption, as during the day time or in a progressively drying soil, water stress develops. In the former case i.e., day time stress, leaves regain turgiscence at night when transpiration demand is reduced and water loss equals water absorbed. But under acute stress plants fail to regain turgiscence at night and reach PWP stage. Arbitrarily the soil water potential at which permanent wilting occurs is taken as -1.5 MPa but in reality it varies widely depending on the crop growth stage and other factors like duration and intensity of the soil moisture deficits and the previous stress history of the crop.

Plants respond to soil moisture deficits in different ways; these include :

(A) *Morphological responses;*

(i) leaf shedding,

(ii) changes in leaf angle, and

(iii) leaf rolling,

(B) *Physiological responses;*

(i) cuticular waxy coating,

(ii) decreased cell division and elongation,

(iii) decreased leaf enlargement/elongation,

(iv) reduced vessel size in roots,

(v) increased root growth,

(vi) stomatal closure,

(vii) rise in canopy temperature,

(viii) osmotic adjustment,

(ix) reduced photosynthesis, and

(x) translocation of photosynthates, and

(C) *Metabolic responses;*

(i) increased proteolysis,

(ii) decreased protein synthesis,

(iii) proline accumulation,

(iv) betaine accumulation,

(v) abscisic acid accumulation,

(vi) decreased cytokinin, and

(vii) decreased activity of nitrate reductase.

In determinate crops like cereals in which yield is produced in a terminal inflorescence, the critical stages are shooting, heading and flowering. In these crops water stress at inflorescence emergence reduced the potential grain setting through reduced fertilization. Likewise water stress at grain filling stage may reduce yield through poor filling of the grains. In indeterminate crops like legumes, oilseed etc. fruits/seeds are borne on lateral inflorescence and the apical meristem does not become reproductive. Both vegetative and reproductive growth continue concomittently. Stress during flowering and pod development stage has serious detrimental influence on yield. In sandy soils where moisture retention is low and the rainfall is not evenly distributed crops are often subjected to repeated droughts. Under such situations, the adverse effects are cumulative. For example, in pearl millet repeated mild droughts at flowering and grain filling stages were more detrimental for crop yield than single intense drought at either of these stages (Lahiri, 1988). In sesame, mild drought at less sensitive seedling stage (20 days after sowing) lessened the adverse effects of crop growth and yield of severe drought at subsequent critical stage. Thus it appears that both complexion and intensity of drought may modulate the crop performance. Sensitive growth stages of some crops are :

Bean : flowering, pod setting

Cowpea : flowering, pod filling

Groundnut : flowering, seed development

Corn : silking, early milk, 12-leaf to blister kernel, flowering

Sorghum : boot to bloom, flowering, grain formation

Wheat : jointing, flowering, crown root initiation

PLANT ADAPTATIONS

Root growth : Trees maintain high rate of water use essentially due to their deep and extensive root system. Root systems of *Acacia senegal, Prosopis cineraria, Tecomella undulata, Albizzia lebbek* possess, apart from tap roots that go deep down in the soil, extensive lateral roots. The roots of *Albizzia lebbek* may spread horizontally upto 20 m (Muthana et al., 1984). The shallow roots extract soil moisture during the wet period from the upper moist soil while deep roots extract moisture from the deeper layers. The desertic grasses also posses extensive root system. For example in *Lasiurus sindicus* 24 per cent of the roots go beyond 1.5 m depth (Chakravarty, 1968). Thus these extensive root systems help plants in maintaining favourable tissue moisture.

Plant water potential : The trees often show only small changes in the leaf turgiscence at various levels of plant water potential. In *P. cineraria*, RT and plant water potential was 90.2 per cent and -0.6 MPa after 31.3 mm precipitation (Table 28). After six days, while plant water potential declined to -2.2 MPa but leaves retained 84.1 per cent RT. Similar trend was observed in *Eucalyptus camaldulensis* (Lahiri, 1988).

TABLE 28. RT (%) and plant water potential (MP_a) in desertic trees a day and six days (within parenthesis) after 31.3 mm shower

Trees	RT	Water potential
P. cineraria	90.2 (84.1)	-0.6 (-2.2)
E. camaldulensis	90.5 (80.6)	-0.8 (-3.3)

Foliar absorption of moisture : Certain plants like *P. cineraria* have capacity to absorb moisture during light showers (Bhatt and Lahiri, 1969). This helps the plant in maintaining favourable tissue moisture for various metabolic activities.

CONCLUSIONS

Thus the desertic trees and grasses possess ability to avoid development of severe stress either due to extensive root system or ability to curtail water turnover through regulation of leaf area. Crops, however, can experience stress of varying intensities and magnitude depending on the crop growth stage. Further work is required to be done on the competition of trees with crops in the initial stages of tree establishment when both tree and grass components may exploit moisture from the same zone.

REFERENCES

Bhatt, P.N. and A.N. Lahiri. 1964. Some evidence of foliar absorption of water in a xeric tree, *Prosopis cineraria*. Naturwiss. 14 : 341-342.

Burman, U., S. Kathju, B.K. Garg and A.N. Lahiri. 1991. Water management of transplanted seedlings of *Azadirachta indica* in arid areas. Forest Ecol. Managem. 40 : 51-63.

Chakravarty, A.K. 1968. Role of some perennial grasses of western Rajasthan in development of grasslands of the arid zone of India. In : Proc. Symp. Adv. Trop. Ecol., p. 620-630.

Lahiri, A.N. 1965. Some aspects of soil water relationship in the growth of plants in arid environment. In : Growth and Development of Plants (R.D. Asana and K.K. Nanda, Eds.), p. 152-168, Today & Tomorrow Book Co., New Delhi.

Lahiri, A.N. 1975. Water use by desertic vegetation and its hydrological implications. Ann. Arid Zone 14 : 135-148.

Lahiri, A.N. 1988. Adaptations of arid zone plants to soil water deficit. In : Desert Biology (I. Prakash, Ed.), p. 111-121, Scientific Publishers, Jodhpur.

Lahiri, A.N. and B.C. Kharbanda. 1966. Studies on plant water relationship. II. Influence of soil moisture on the transpiration of *Tecomella undulata* seedlings. Proc. Nat. Inst Sci., India 32 : 34-46.

Lahiri, A.N. and V. Kumar. 1967. The annual water turnover from a xeric tree, *Prosopis cineraria*. Sci. Cult. 33 : 77-78.

Muthana, K.D., G.L. Meena, W.S. Bhatt and O.P. Bhati. 1984. Root system of desert tree species. My Forest 20 : 27-38.

13

Soil Fertility Management in Agroforestry Systems

R.K. Aggarwal and Praveen-Kumar

The trees are essential component of agroforestry system. Their utility in this system is due to their roots which move deep exploiting large soil volume often inaccessible by plant roots. A part of nutrients thus exploited is deposited back on the surface through litter fall. This process is referred as albeit erroneously nutrient pumping by trees. An increase in the availability of N, P and K (Aggarwal and Praveen-Kumar, 1990) and crop growth (Aggarwal *et al.*, 1993) beneath the *P. cineraria* as compared to the adjacent area has been observed (Tables 29 and 30).

TABLE 29. Chemical characteristics of surface soil (0-15 cm) under *P. cineraria* (khejri) and adjacent open field (non- khejri)

Characteristics	Non-khejri soil	Khejri soil
pH	8.20	8.00
OM (%)	0.37	0.57
N (%)	0.02	0.04
S (%)	0.02	0.03
P (%)	0.03	0.04
Available nutrient (kg ha^{-1})		
N	190.00	250.00
P	7.7	22.4
K	370.00	633.00

Source : Aggarwal and Praveen-Kumar (1990)

TABLE 30. Dry matter yield (g pot^{-1}) of pearl millet in *khejri* and non-khejri soil

Treatment	Non-khejri soil	Khejri soil
Check	17.4	52.6
NPKS	39.5	71.0
NPK	28.9	70.9
NKS	30.7	66.6
NPS	33.4	69.0
PKS	25.8	60.3

Source : Aggarwal et al. (1993)

Such phenomenon of micro site enrichment by some woody perennial species is extremely important in agroforestry. Besides, one of the advantage that is most commonly attributed to agroforestry is the possibility for improvement of fertility status of agricultural lands through additional amounts of nitrogen that could be added to the system by the tree legume component. Although legumes are not the only group of woody species that have a potential role in agroforestry, the family Leguminosae offers by far the greatest choice of woody species for agroforestry in terms of economic uses as well as ecological adaptability. Di-nitrogen fixation is not a natural phenomenon that is universally present in all members of the Leguminosae family. Out of the three sub families of Leguminosae, the majority of species in Mimosoideae and Papilionoideae are known to fix nitrogen, in contrast to only a smaller percentage of the members of Caesalpinioideae. According to the estimates of Allen and Allen (1961), there are about a thousand species of tropical woody legumes which form di-nitrogen fixing root nodules. A few non leguminous tree genera can also fix nitrogen, notably the tropical genus *Casurina*.

FERTILITY MANAGEMENT IN DIFFERENT LAND USE SYSTEMS RELATED TO AGROFORESTRY

Nair (1980 and 1983) has described various types of agroforestry systems, prevalent in the S.E. Asia and arid and semi arid regions (Table 31). However, shifting cultivation, taungya system, agricultural plantation crops are the dominant systems which play significance role in fertility management.

TABLE 31. Prominent agroforestry systems and practices in the S.E. Asia and Arid and Semiarid regions

Prominent systems	S.E. Asia	Arid and Semi-arid regions
Agri-silvi-culture	1. Commercial trees among crops	1. Use of trees on farmlands for protective role (wind-breaks, dune fixation)
	2. Fruit/shade trees among crops	2. Productive + protective role of trees on farms (*A. albida* / *Leucaena*) + agric. crop systems
	3. Live fences	
	4. Shelterbelts	
	5. Taungya	
	6. Shifting cultivation systems	
	7. Intercropping in plantation crops (rubber, oil palm, coconut)	
Silvi-pastoral	1. Pasture in forest plantation	1. Nomadic (semi-nomadic/ transhuman)
	2. Pasture in secondary forests	2. Sedentary livestock grazing systems/browsing systems
	3. Commercial trees in pastures	3. Fodder tree/shrub systems
	4. Fruit/shade trees in pasture	
	5. Fodder trees	
	6. Coconut + pasture	
Agri-silvi-pastoral	1. Crops and grazing in plantations	1. Forestry dominating (forest lands)
	2. Agri. tree crops + grazing in forest plantation	2. Agriculture dominating (crop lands)
	3. Multipurpose trees with crops/animals	
	4. Integrated farming systems with agric. plantation crops (rubber, coconut, oil palm)	3. Livestock dominating (rangelands)

Home gardens	Various forms of multi-species combination	Various forms
Others	1. Silviculture in mangrove forests	1. Oasis
	2. Agri-silvi-fishery	2. Irrigation systems
	3. Trees on bunds in fish breeding ponds	3. Various site-specific systems
	4. Swidden farming	
	5. Fuelwood agroforestry	

Shifting Cultivation

Shifting cultivation is a land use system in which forest land is cleared and cropped for few years, and then left for regeneration of bush (fallow period), the duration of fallow period being much longer than cropping period. Clearing of the forest, disrupts the close nutrient cycle which exists in the soil forest system leading to increased availability of nutrients. Therefore, first crop grown after clearing the forest is benefited significantly. Afterwards, the soil becomes less and less productive.

Various improvements on shifting cultivation have been proposed for improving fertility status of such soils during the fallow period. Such alternatives include the corridor system in Zair (Jurion and Henry, 1969), improved cropping system in India (Borthakur et al., 1981) and the use of fertilizers and other modern methods of farming.

The Taungya System

In this system the crops are grown together with the forest plantations in the early stages of their establishment. Although the wood production is the ultimate objective, the immediate motivation for practising it is food production.

In taungya system, the soil fertility may decrease mainly due to the possibility of erosion, during the preparation for planting the crops. In high rainfall areas, the surface horizons become partly eroded and sub surface horizons may get gradually exposed. The addition of crop residues under such conditions can help in restoring soil fertility and minimizing the erosion hazard.

Woody Perennial Plantation

A sizeable percentage of total production of most of the

plantation crops is produced in small holdings. These form significant proportion in terms of both population they support and the area they cover. In these small holdings, the farmers usually integrate crops and animal production with perennial crops primarily to meet their food requirement. A common example is growing of leguminous cover crop with the commercial plantation of rubber and oil palm. A recognized advantage of this is the enrichment of soil through di-nitrogen fixation. However, the application of manures and fertilizers and other soil management considerations are important if reasonable yields of interplanted crops and main (plantation) crop is to be ensured.

CONCLUSIONS

On the basis of above, it can be concluded that nutrients can be more efficiently used under the agroforestry system. However, to maintain a higher level of fertility a number of factors, including the environmental conditions, tree and crop characteristics and soil conditions and above all, management practices must be looked into.

REFERENCES

ggarwal, R.K. and Praveen-Kumar. 1990. Nitrogen response to pearl millet (*Pennisetum typhoides* S&H) grown underneath *Prosopis cineraria* and adjacent open site in an arid environment. Ann. Arid Zone 29 : 289-293.

Aggarwal, R.K., Praveen-Kumar and P. Raina. 1993. Nutrient availability from sandy soils underneath *Prosopis cineraria* (Linn. Macbride) compared to adjacent open site in an arid environment. Indian For. 19 : 321-325.

Allen, E.K. and E.O. Allen. 1961. The scope of nodulation in Leguminosae. Recent Adv. Bot. 1 : 585-588.

Borthakur, B.N., R.N. Prasad, S.P. Ghosh, A. Singh, R.P. Awasthi, R.N. Rai, A. Verma, H.H. Datta, J.N. Sachan and M.D. Singh. 1981. Agroforestry based farming system as an alternative to *Jhuming*. In : Proc. Agroforestry Seminar, Indian Council of Agric. Research, New Delhi, p. 109-131.

Jurion, F. and J. Henry. 1969. Can primitive farming be modernized ? INEAC Ser. HORS. Brussels, Institut National pour L'Etude Agronomique du Congo, p. 445.

Nair, P.K.R. 1980. Agroforestry species. A crop sheet manual. Nairobi, ICRAF, p. 336.

Nair, P.K.R. 1983. Tree integration on farmland for sustained productivity of small holdings. In : Environmentally Sound Agriculture, Lockeretz, W. (Ed.) Praeger, New York, p. 333- 350.

14

Feed Resources of Indian Arid Zone : Their Nutritive Value for the Livestock

H.C. Bohra and B.K. Mathur

Fodder resources in western Rajsthan are scanty, and most of the feeding stuffs remain nutritionally very poor except during monsoon. The local desert breeds of livestock, however, have acquired certain characteristics which help them in sustaining harsh arid climatic conditions. These characcers have been transmitted through generations and resulted into evolution of hardiest breeds of the livestock. But, due to insufficient supply of the nutrients required by the animals, their genetic potentiality could not be expressed to their optimum production level, hence, animals produce less than their actual level of production. Productivity of desert animals can successfully be augmented by improving their nutritional status.

Feeds are the source of nutrients required by the animals for performing various body functions, growth and for synthesizing the products. Broadly, livestock feeds of this region can be classified as follows.

CONCENTRATE FEEDS

These are principal crop products and rich in digestible nutrients. Some of them are rich in protein and/or lipid. *Bajra* (*Pennisetum typhoides*), barley (*Hordeum vulgare*), *jowar* (*Sorghum bicolor*), wheat (*Triticum vulgare*), Maize (*Zea mays*) are carbohydrate rich grains. *Guar* (*Gymopsis tetragonoloba)* and moth *(Phaseolus acontifolius*) are good source of protein while cotton seed and *til* (*Sesamum indicum*) are rich in both lipid and protein.

Although the feed stuffs of this class are most nutritious amongst all the feeding stuffs, due to uncertainties of rainfall and frequent droughts occurring in the region, their availability is not assured. Further, as most of these feed stuffs are staple food for the human population of the region, their availability for livestock consumption is limited. The guar grains which were previously, extensively and exclusively used for the livestock feeding is now used for gum production. Even protein rich by-product of guar-gum industry is used in confectioneries and, thus became unavailable for the livestock consumption. Oil seeds are not used as such for livestock feeding but their cakes such as cotton seed, groundnut and til are rich in crude protein and used extensively for feeding productive animals. Amongst the oil seed crops only til is grown in the area, but the production is too low to fulfil the requirement. Recently a technique has been developed at CAZRI to feed/locally available nonconventional feed, *Tumba* (*Citrullus colocynthis*) cake, a waste from the oil industry, for feeding the productive cattle. Before feeding, the cake is to be soaked in water to remove its bitterness which is due to citrullin content. Use of *Tumba* cake has been able to replace 25 per cent of the conventional concentrate mixture.

PASTURE HERBAGES (INCLUDING GRASSES)

Desert flora is mainly composed of annual forage, though perennials are also grown in a limited area. During monsoon, naturally growing lush green forages, mainly annuals, are available in abundance. But, as the monsoon withdraws, these species are either consumed by the grazing animals or dry up and their nutritional quality deteriorates. Nutrient contents of even promising indigenous perennial grasses decline repidly with the advancing age. For instance 7.7 per cent crude protein content of *Cenchrus ciliaris* at preflowering stage reduces to 2.7 per cent at maturity.

AGRICULTURAL BYPRODUCTS

Diet of desert livestock mainly consists of dry grasses and straw of agricultural crops. The feeding stuffs are generally low in digestible nutrients containing less than 1 per cent digestible crude protein. Due to high crude fibre and low total digestible nutrients, most of the energy provided by these feeds to the animal is concumed in mastication, digestion and assimilation, hence, the net amount of the energy available to animal is very low. Under the green revolution programme, plant breeders have evolved high grain yielding varieties of food crops. This has resulted in reduction in the production of nutritionally poor straw. Availability of nutrients from these feeds could successfully be improved by treating with calcium hydroxide; spraying of sodium hydroxide solution; and by urea mollasses treatments. Recently, techniques have been developed for nitrogen enrichment of these feeds by treating the poor quality roughages with urea and anhydrous ammonia. Studies conducted at IVRI, Izatnagar and NDRI, Karnal have shown that wheat straw and sorghum stover combined with urea and molasses can not only meet the maintenance requirement of cattle, but part of the production also.

A technique has been developed at Animal Microbiology Laboratory at CAZRI, Jodhpur for improvement of poor quality feeds like grasses and *bajra* straw. These fodders are ensitled (60 days) with urea (1 per cent) and molasses (8 per cent) and in different combinations with other fodders to increase the protein and carbohydrate values. The process has been further modified to suit arid conditions by adding lactic acid bacterial culture through a simple and cheap source-buttermilk, since the natural epiphytic flora of these bacteria is low in arid regions. By this modified ensiling technique, grasses and straws can be preserved with improved nutritive value and can be used during lean periods.

TOP FEEDS OR TREE LEAVES

Tree leaves, though have limited importance as livestock feeds in other parts of the country, in arid regions of Rajasthan where availability of agricultural products and byproducts is uncertain, remain best choice for feeding the livestock. The top feeds are generally rich in preformed water, crude protein and calcium; but low in less digestible nutrient, the crude fibre. Unlike other protein rich feeds, the top feeds do not cause bloat in the animals, even if the animals are solely fed on the tree leaves. Fresh leaves contain as such as 60 to 70 per cent preformed water. The animals fed with such leaves can get considerable quantity of water which, inturn, reduces free intake of water by these animals. Unlike desert forages, the nutrients level in these feeds, remain nearly same throughout the year.These feeds can safely be used as maintenance ration, and as supplement to less productive animals maintained on dry grasses and straws. *khejri* (*Prosopis cineraria*) and *bordi* (*Ziziphus nummularia*) are most promising top feed species of this region. Tree species such as *Neem* (*Azadirachta indica*), *Babool* (*Acacia nilotica*), *Pipal* (*Ficus religiosa*), *Bargad* (*F. benalensis)* and *Ardu* (*Ailanthus excelsa*) are also important top fed of the region. These trees provide appreciable quantity of leaves and can contribute considerable amount of nutrients to the desert livestock. On dry matter basis, the *Ardu* leaves contain highest level of cell contents and most nutritious among the desert tree leaves consumed by the animals. High contents of digestible crude protein (DCP) and total digestible nutrients (TDN) of ardu leaves further indicate its importance as animal feed. Thus, the propogation of plants having more fodder yields with better nutritive values can certainly be helpful in improving the nutritional status of the livestock population of the arid region of Rajasthan.

However, presence of anti-nutritional factors in some of these feeds adversely affects the availability of nutrients to the animals. Leaves of *khejri* contain high level of tannins while that of *Bordi* have high lignin content. The tannins combine with dietary or microbial proteins and form conjungate which is non- degradable in the hind gut of the animals. The lignins, on the other hand, make the cell wall constituents (especially the hemi-celluloses) undigestible. Chemical treatments have been tried to remove or inactivate these antinutritional factors. Protein availability from *Pala* leaves has successfully been improved by formaldehyde treatment. Tannins of *khejri* leaves can be inactivated by treating with ferric chloride solution. Recently a technique at CAZRI has been evolved for removing tannins from *khejri* leaves by soaking in 0.5 N sodium carbonate solution (5.3 per cent aqueous) for 24 hours, following washing with water. This can remove about 97 per cent of total tannins present in raw *khejri* leaves.

CONCLUSIONS

Erratic rains and frequent droughts experienced in arid areas adversely affects the availability of crop products for the livestock of this region. The large scale use of chemically or otherwise treated agricultural by-products and tree leaves, as animal feeds can improve the nutritional status and thus, the productivity of the livestock of the region.

15

Potential Role of Agroforestry in Arid Zone in 2000 AD

B.M. Sharma and J.P. Gupta

The growing demand for food and fodder due to escalation of human and animal population and shrinkage of available land resources necessitate to increase the production per unit area. The task is far more challenging in arid regions where the availablity of good quality land and water is scarce. Besides, the high input technology to boost the production such as monocropping with improved crop varieties, using irrigation and high level fertilizers and pesticides,etc. are neither available to arid zone farmer nor he can use due to environmental constraints. This is primarily due to infertile erosive soils, insufficient and erratic rainfall and marginal economy. However, agroforestry, by virtue of incorporation of perennials and herbaceous plants in the same land management unit, can provide multitude of products and services, can help the resource poor farmers to improve productivity and meet the future demand for fuel, fodder, food timber etc.

FUELWOOD AND FODDER PRODUCTION

Wood products such as fuel, timber and poles for building, fencing and for wide variety of purposes in farm and home construction are important components of the everyday life of rural people. According to an estimate the total fuel wood requirement in the country is 143 m tonnes while the total production is only 49 m tonnes. In arid areas, 2.4m tonnes fuel is required, which is expected to increase to 3.6 m tonnes by 2000 AD. In order to meet the additional demand plantation of more trees on village community lands, field boundaries and other waste lands is essential. *Acacia tortilis, Prosopis juliflora, Azadirachta indica* are suitable tree species for this purpose. Plantation of trees on crop lands is another approach to meet the fuel demand as the availability of such lands is also limited. However, the trees and crops selected for this purpose should be complementry to each other. By this approach most of the rural needs can be met from the farm itself rather than have to be purchased with limited financial resources.

Several plant species are important from medicine and industry point of view. *A. senegal* and *P. juliflora* produce commercial quality gum. *Calotropis procera* and *Euphorbia* sp. which can be grown even on marginal lands, are promising source of hydrocarbons. Some minor products such as resins, gum, fibre, rubber, honey etc. can also be obtained from trees. The importance of multipurpose trees and shrubs in low inputs farming systems has been highlighted by Swaminathan (1987).

The demand for fodder has increased considerably over the past few years while the supply has declined. Based on the net balance of demand and supply for fodder at 1992-93 level of livestock population there was a deficit of 16.9 m tonnes in arid region of Rajasthan.The livestock population in 2000 AD expected to be 11.3 m ACU (adult cattle units) i.e. 17.6 per cent more than 1983 level and accordingly, by 2000 AD the demand for fodder would be 28.2 m tonnes. The increased demand for fodder can be met by putting more areas under grazing lands, planting fodder trees, adoption of silvipasture in low rainfall areas and degraded lands and the improved management practices. *P. cineraria, Z. nummularia, T. undulata, A. lebbek,* and *A. senegal* are suitable tree species for arid regions. Grasses like *Cenchrus ciliaris,Lasiurus sindicus, Dichanthium annulatum, D. antidotale* are compatible with these tree species. *A. tortilis* alone can provide 6000 kg ha^{-1} fuel wood and *C. ciliaris* alone can provide 4800 kg ha^{-2} forage while a combination of the two can provide 5000 kg ha-1 fuel wood and 5580 kg ha^{-2} of forage. *Z. rotundifolia* has been found to have compatibility with *C. ciliaris* and *L. sindicus* under varying rainfall conditions over eight years under arid conditions (Sharma et al.,1993). Venkateswarlu (1989) estimated the forage production under different tree species of arid Rajasthan and observed that *P. cineraria* and *T. undulata* can provide 1500 kg ha^{-1} forage (Table 32).

TABLE 32. Forage production under different trees in Arid Rajasthan

Tree species	Total forage kg ha^{-1}	% grass
P. cineraria	1500	74.8
T.undulata	1500	53.7
A.lebbek	1400	57.9
Zisiphus sp.	1000	87.5
Acacia senegal	600	52.5

SUPPLEMENT TO FERTILIZER NEEDS

According to an estimate (Gaur and Singh, 1982) the total nutrient requirement in 1989-90 was 22.1 m tonnes and half of it was met from the inorganic fertilizers and remaining from other sources (Table 33). By 2000 AD this requirement would increase to 28.8 m tonnes out of which, as per the existing production in the country, 75 per cent can be met from inorganic fertilizers. As such 25 per cent of the total requirement has to be met from some other sources.The existing levels of 10 to 40 kg ha^{-1}N, 9 to 20 kg ha^{-1} available P and 350 kg ha^{-1} K, though adequate for the production of natural vegetation in arid areas, addition of nutrients from extra source is essential.

TABLE 33. Estimated nutrient requirement in the country (in tonnes)

	1989-90	1995-96	1999-2000
Nutrient requirement	22.1	25.1	28.8
Fertilizer consumption	11.9	16.3	21.6
Contribution from soil, organic and biological sources	10.2	8.8	7.2

Source : Gaur and Singh (1982).

The availability of manure is very low in arid areas as most of it is used as fuel. Therefore, nutrient addition to soil through natural vegetation can significantly supplement the fertilizer requirement. Presence of trees or shrubs in the crop fields can add leaf litter and improve the fertility of soil. Addition of 69, 39.2, 23.2, 30.3 and 23 g m^{-1} leaf litter through *Prosopis cineraria,Tecomella undulata,*

Albizzia lebbek, Acacia senegal and *Prosopis juliflora* was reported by Venkateswarlu (1989). Inclusion of leguminous trees in the agroforestry systems add nitrogen to soil through fixation. One year old seedlings of *Acacia nilotica, Hardwickia binata, Colophospermum mopane* and *Dichrystus nutans* grown alongwith crop can fix 10 to 15 kg N ha^{-1} (Rao, 1992).

CONSERVATION OF NATURAL RESOURCES

Conservation of resources on which production depends is most important component of sustainability.In arid regions, because of the high wind speeds a considerable portion of the top fertile soil is lost through erosion which also carries nutrients and organic matter with consequent lowering of the crop yields. The situation is aggravated as water is also scarce.

There are two roles of the tree component in erosion control : supplementary and direct. In the supplementary role, erosion control is achieved by conventional methods such as earthbanks and ditches, terrace risers or grass strips. The trees serve to stabilise the conservation structures or to make the productive use of land they occupy. In the direct role, the trees themselves are responsible for erosion control which can be achieved in two ways with trees acting as barriers or as vegetation cover. The barrier function is by reducing wind speed, checking runoff and suspended sediments. The cover function involves reducing rain drop impact and wind action on soil particles by increasing soil cover, with living or dead plant materials. Mixtures of trees and shrubs planted across the wind direction helps in reducing the wind speed,evaporative demand of the atmosphere and wind erosion. Shelterbelts and micro shelterbelts play significant role in reducing the erosion. Grasses not only improves the soil fertility but also have sand binding effect and thus control the erosion losses.

Soils of the arid zone are predominantly light textured. Therefore about 30 to 40 per cent of the rainwater goes deep into the murrum substratum beyond the root zone. and becomes unavilable to most crop plants. Tree companent of agroforestry system, because of the deep root system can efficiently utilize this water. *Prosopis cineraria* is one such tree of arid zone. Water infiltration in the soil is increased by the tree root action and the accumulation of absorbent humus on the soil surface thereby significantly reducing the volume, velocity and erosivity of surface runoff. The reduction in evaporation losses due to decrease in incoming radiation to the soil surface and soil temperature. Organic matter, nutrients and profile moisture storage under tree crop/grass combination was higher as compared to tree alone. (Sharma et al.,

1993, Sharma et al., 1994). Thus agroforestry maintain more favourable physical properties than monocropping, through a combination of organic matter maintenance and the effects of tree roots.

Tree root systems intercept, absorb and recycle nutrients in the soil solution that would otherwise have been lost in leaching, so making the nutrient cycle more closed. The mycorrhizal systems associated with the tree roots are agents in this process through their penetration of large proportion of the soil volume leading to uptake of nutrients which can move only short distance by diffusion.

RECLAMATION OF DEGRADED LANDS

Reclamation of lands degraded due to erosion or infestation of salts is one of the important aspects of increasing production. At the early stages of reclamation, severely eroded soils require a vegetation cover coupled with protection from grazing. Trees, offers an opportunity of better coverage of soil surface. Once the soil fertility is sufficiently restored, this can be converted to carefully protected silvipasture or agripasture lands.

Plantation of some forest trees on the alkali soil has been found very effective in reducing the pH and improving physical properties of soil (Yadav and Sharma, 1977). This can be combined with progresive introduction of cropping. While tree root systems are undoubtedly of value in such reclamation, they are of no use unless associated with standard reclamation measures, such as gypsum treatment.

SUSTAINABILITY AND ECONOMIC STABILITY

Agroforestry provide sustainability by maintaining the production levels alongwith conservation of resources. The human population in the arid areas is expected to increase from 16.16 million in 1991 to 19.81 million by 2000 A.D. The increased demand can be met through the use of high yielding crop varieties, use of fertilizers and pesticides and water management practices. Introduction of grasses and trees in the cultivated lands is a measure which improve crop production. Studies have indicated that 40 to 70 trees ha^{-1} in Jodhpur region and 104 trees ha^{-1} in Sikar region has no adverse effect on understorey crops (Gupta and Sharma, 1989). Similarly, *Z. rotundifolia* planted at 5x5 m spacing had only marginal effects on the yields of pearlmillet, clusterbean (Sharma et al., 1994) and grasses (Sharma et al., 1993). Besides,agroforestry has following important role in rural areas.

It provide greater insurance against weather abnormalities. Many multipurpose trees such as *Prosopis cineraria*, *Z. rotundifolia* and *T. undulata* thrive well in arid areas. The tree component helps in utilization of off season rainfall. As most of the drylands are cropped for one season, the precipitation occurring during off season go unutilized, but presence of trees in the system can make use of the same efficiently. During off season there would be no field work for family labour in arid areas. This time can be effectively used for harvesting/pruning/marketing of tree components. Presence of trees in the system gives extended management options for the farmers i.e. the trees can be harvested either for fodder, fuel, pole or small timbers as per farmers requirement and market demand.

SELECTION OF TREE SPECIES FOR AGROFORESTRY SYSTEMS : SOME CONSIDERATIONS

To reach a decision with regard to the species to be planted or favoured for agroforestry it is necessary to consider the requirement of possible species in relation to any special features of the site within the general climate, soil and vegetation type of the region. It is also important to consider the mutual compatibility and non-competitiveness of various agricultural and tree and/or shrub species to be introduced. Special regard should be given to avoiding species that pose problems of competition for water, nutrients and solar energy. The desirable characteristics of an agroforestry tree can be as follows :

(i) Considering the important and varied roles that trees can, and are expected to play in the agroforestry, it becomes imperative that the trees be multipurposes one. The trees should provide leaves, fruits, flowers and wood to meet the various needs of the farmers.

(ii) Deep rooted so that it can effectively stablize the soil, so its roots do not compete for moisture and nutrients with the shallow roots of annual food crops, and so it can reach lower water table and surviving during dry periods.

(iii) Light-crowned so it does not totally shade intercropped food plants.

(iv) Rapid rate of growth to reduce waiting period for tree products by farmers.

(v) Self-pruning properties or if not self pruning, it should be able to tolerate relatively high intensity of prunning.

(vi) Its phenology, particularly with respect to leaf flushing and leaf fall, should be advantageous to the growth of annual crops.

(vii) It should be tolerant of side shade.

(viii) It's rate of litter fall and litter decomposition should have positive effect on soil.

(ix) It should be nitrogen-fixer so that it can contribute to the rehabilitation/improvement of the soil.

(x) It should be good coppier so that it can regrow after periodic cutting without requiring replanting.

(xi) Low requirement for water and nutrients.

REFERENCES

Gaur, A.C. and Ramender Singh. 1982. Integrated nutrient supply system. Fert. News 27 : 87-98.

Gupta, J.P. and B.M. Sharma. 1989. Annual Report, CAZRI, Jodhpur.

Sharma, B.M., S.S. Rathore, J.P. Gupta and R.N. Prasad. 1993. Evaluation of some silvipasture systems under arid conditions of Rajasthan. J. Trop. Forestry 9 : 103-109.

Sharma, B.M., S.S. Rathore and J.P. Gupta. 1994. Compatibility studies on *Acacia tortilis* and *Ziziphus rotundifolia* with field crops in arid conditions. Indian Fer. 120. 423-429.

Swaminathan, M.S. 1987. Agroforestry in Asia, Anniversary Celebration, ICRAF, Nairobi.

Venkateswarlu, J. 1989. Strategies for harvesting natural resources in arid west Rajasthan. Indian Fmg. 39 : 19-20 and 31.

Yadav, J.S.P. and B.M. Sharma. 1977. Soil changes in differently treated planting pits under Eucalypus hybrid in a saline sodic soil, Agrokem Talajt, 25 : 327-329.

16

Insect Pest Management in Agroforestry

M.P. Singh

In an agroforestry field where the herbaceous and woody plants are co-cultivated, there are interactions among them,the intensity of which depends on the type of the system and species involved. The associated insects react to these interactions and thus the insect pest situation in an agroforestry system may differ from the monoculture plantation where inter species competition is inoperative (Rathore, 1995).

The insect plant relationship in agroforestry systems ranges between the two extremes of usefulness and harmfulness. Introduction of woody plants in an agricultural or pastoral field may change the pest situation; and so does the augmentation of herbaceous plants in woods. From a pest management point of view, there could be many advantages in co-cultivation of selected crop/grass/woody plants. It may be possible through proper selection of species, field designing and plant sequencing etc, to add to the sustainability of the agroforestry system through minimization of insect pest losses on a long term basis (Altieri *et al.*, 1983).

SPECIES SELECTION

The type of plant species in an agroforestry system has a bearing on the insect populations (Kareiva, 1983). An increase in the availability of host plants in a system may lead to increased insect activity. In the arid regions, the choice of woody species is rather limited. In such cases, it is the crop component and sequencing that could be manipulated in order to avoid round the year availability of food plants for the insect pests. Legume-legume combinations or sequencing may provide opportunity to the insects to multiply incessantly. A non-host companion plant on the other hand, will serve to limit pest multiplication. In the systems involving *Albizia, Prosopis, Acacia* or *Cassia* spp. as woody componets and cowpea, mung bean, moth bean, cluster bean or groundnut as crop components, the likelihood of bruchid infestation is increased, as all the plants involved are host to this group of insects. Likewise, the other legume pests such as jassids and bugs may also thrive well. Therefore, for such systems, it is essential to provide preventive or curative measures. Crops like millets may be considered as an alternative of pulses, the woody plants of choice could be jujube or other non legumenous tree spp.

SPECIES DIVERSITY

From a pest management point of view, an agroforestry system involving greater number of plant species may be more sustainable than the one which consists of fewer plant species. However, pest levels may not be reduced to the same degree in polyculture systems (Risch *et al.*, 1983) still, the non-related species shall make the system but more sustainable. In an agroforestry system with greater species diversity, the pests specific to plant species could easily be checked. Other pests with limited host range could also be checked from uncontrolled multiplication. In fact, subject to availability of information about the key pests of an area, it should be possible to raise an agroforestry system consisting of the non-host plants for these pests.

Presence of non-host plant species along with the host plants also contributes towards reduction in the activity of insect pests (Saxena and Basit, 1982). These plants may act as barriers for the movement of pests. Therefore, deliberate inclusion of such non-host plants in the system will lead to reduction in pest problems. Conversely, raising of highly prefered plants just to serve as trap for the pests of the main crops, may also yield similar outcome. The trap plants will attract the pests where these could be destroyed, without touching the main crops/ woody plants.

Allowing the non crop plants to survive around the fileds is another way of increasing the biodiversity in the agroforestry systems. In fact many of the insect pest problems in the past have been created through indiscriminate destruction of non crop plants from around the crop fields. Only the obnoxious weeds should be removed, and those plants which do not harm the main crops or offer negligible competition, should be allowed to exist.

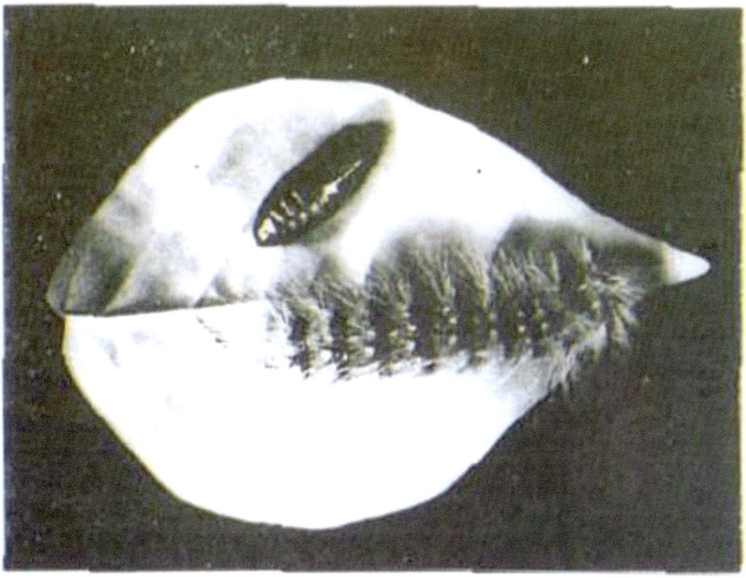

Fig. 6. Red Hairy Caterpillar, the most dreaded pest of *Kharif* crops.

Fig. 7. Chickpea pod Borer.

FIELD CONFIGURATION AND DESIGN

The configuration of a land unit is determined by the plant composition, their colour, structure, height, density in the field and the type and colour of the background soil. Change in field configuration affects the orientation by insects towards the field. Dempster and Coaker (1974) observed that even small changes in cropping practices could alter substantially the attractiveness of the field to the pests. Through proper design of an agroforestry system, configurational changes could be effected, leading to reduced colonization by the pests. Crop and woody components could be raised in a way to take the best advantage of sun and wind direction. The effect of shading should be minimal in systems where trees have been planted along sun's path. The effect of weather would be the least in systems with thick vertical and horizontal vegetation around the boundary. The size and density of plants and their relative arrangement in time and space will determine the sphere of activity of the pests and their natural enemies.

MICROCLIMATIC CHANGES

Introduction of tree components in a grass/crop field or vice versa, brings about changes in radiation penetration,temperature and humidity profile, transpiration rate and soil temperature etc. A mature system may also be capable of effecting wind regimes and rainfall interception. All these factors in turn affect the activity of insects-both pests and their natural enemies. Whereas the shade loving phytophagous insects may find these conditions to grant them longer period of feeding activity, those insects which prefer peripheral sites may have to shift away from shaded areas. Sucking insects as the aphids, jassids, white flies and bugs may find the conditions under agroforestry fields conducive to multiply faster or inflict greater injury than in fields directly exposed to the sun.

BIOCONTROL POTENTIAL

The presence of perennial woody plants in agroforestry systems may prove quite advantageous from insect pest management angle. These plants play an important role both when crops are in field and also during the off season and during the period between two successive crops. The woody plants provide shelter, food and breeding sites for the natural enemies of insect pests of crops/grasses/trees. If the woody species happen to be one of the food plants of crop pests, these may also help to maintain the population of pest insects on which the parastites and predators multiply during the absence of the crop from the field.

Most predatory and parasitic insects find shelter on the woody plants in agroforestry fields. The density of spiders may be more in the crops raised under *Prosopis cineraria* trees in comparison to that in the open field in the arid zone of Rajasthan (Chandra, 1987). Several species of wasps could be observed on *Prosopis* and *Acacia* plants during the rainy season and afterwards.

Flowers of some woody plants constitute the source of food for a number of parasitic and predatory insects in their adult stage. A large number of Hymenopteran, Dipteran and Coleopteran insects have been witnessed visiting flowers during blossom in *Prosopis* and *Acacia* species in the semi arid regions of Rajasthan, some of which could be the natural enemies of crop or grass pests. Such woody plants which bear flowers for a longer duration may be more useful in agroforestry system, as these may encourage the activity of the beneficial insects like predators and parasites, besides providing edible material for the honeybees. This may be kept in mind while selecting plants for agroforestry.

CONTROL MEASURES

The vulnerable crops under agroforestry systems have to be guarded against the pests during epidemics. The initial establishment phase of most agroforestry systems is not very different from agrarian or pastoral systems and therefore, is as prone to insect pest attack as these systems. The simpler agroforestry systems in their establishment stage may face greater damage than established complex systems. Polyphagous pests, which feed on a variety of vegetation, pose greater problems to agroforestry system as compared to the specific pests or which have a narrow host range. Pests like the hairy caterpillars (Fig. 6) and locusts (Fig. 9) which appear sporadically, may wipe out the entire foliage. The insect pests common to many plant species in the system (Fig. 7 & 8) have to be tackled on priority. For managing these pests, all available measures ought to be deployed. Besides physical, cultural and chemical methods, microbial control methods may also be applied as there are better chances of success of microbes in acting on the target insects under agroforestry conditions.

To sum up, it could be stated that agroforestry practices could be utilized to direct the system in a way that the pest injury is minimised. Through proper species selection, designing and management, conditions could be developed which are favourable for the natural enemies of the pests while being unfavourable for the latter. Such an agroforestry system will be aiding towards its own long term sustainability, other factors being favourable.

Fig. 8. Pest compex of mungbean in the arid and semi arid areas.

Fig. 9. Desert locust, *Schistocerca gregaria*

TABLE 34. Some likely pests in agroforestry systems in the arid and semi arid regions

Tree/crop	Pest species	Other host plants
Prosopis cineraria	*Acaudaleyrodes rachipora*	*Acacia, Albizzia, Dalbergia, Cassia, Leucaena, Bauhinia*
	Adoretus sp.	Polyphagous on crops, trees
	Anomala sp.	-do-
	Caryedon spp.	Legume trees, pulses, peanut
	Celosterna scabrator	Eucalyptus,many woody trees
	Chrysobothris parvipunctata	many woody tree species
	Cosmogaster sp.	-
	Eriophes prosopidis	-
	Eucosoma bioplintha	-
	Homoeocerus prominulus	-
	Lachnosterna consanguinea	wide range of vegetation incl. trees, crops
	Lobopteromyia prosopidis	-
	Omanocoris versicolor	-
	Oncideres sp.	Legumenous, other trees
	Oxyrachis tarandus	grasses, trees
	Pediobopsis sp.	
	Rhinyptia sp.	polyphagous
	Taragama siva	*Acacia, Tecomella*, tamaring, jujube, others
Albizia leböeck	*Achaea janata*	castor, ber pomegranate
	Bruchidius sp.	beans, pulses, legume trees
	Caryedon spp.	legume trees, groundnut
	Drosicha sp.	ber, guave, mango
	Eurybrachys tomentosa	ber, clusterbean, neem

	Halys dentatus	many tree species
	Indarbela sp.	ber, *Acacia*
	Kerria sp.	ber
	Myllocerus spp.	Kharif crops,grasses,tree
	Oxyrachis tarandus	saplings and seedlings
	Psylla sp.	citrus fruit plants
	Xyleborus sp.	many woody plants
Tecomella undulata	*Aphis craccivora*	Polyphagous
	Patialus tecomella	—
	Poekilocerus pictus	-do-
	Stegmatophora sp.	—
	Taragama siva	Polyphagous
Dalbergia sissoo	*Dichomeris eridantis*	—
	Oxyrachis tarandus	Legume trees, grasses
	Plecoptera reflexa	—
Neem	*Halys* sp.	*Prosopis* spp., *Albizia*
	Holotrichia spp.	wide range of plants
	Pulvinaria maxima	—
	P. azadirachtae	—
Cowpea	*Amsacta moorei*	wide range of vegetation
	Aphis craccivora	pulse and other crops
	Callosobruchus spp.	pulses, legume tree seeds
	Cyrtozemia cognata	mungbean, mothbean, guar
	Empoasca sp.	pulse crops, grasses
	Epilachna sp.	Kharif pulse crops
	Heliothis sp.	many kharif and Rabi crops
	Holotrichia sp.	wide range of plants
	Melanagromyza phaseoli	*Kharif* pulse crops
	Mylabris sp.	*Kharif* crops
	Spodoptera sp.	wide range of crop plants

Chickpea	*Agrotis* sp.	pulse crops
	Callosobruchus sp.	pulse crops, legume trees
	Heliothis sp.	wide range of crop plants
	Spodoptera spp.	-do-
Cluster bean	*Aphis* sp.	cowpea,other pulses
	Bemisia tabaci	wide range of plants
	Conicobruchus sp.	—
	Cyrtozemia cognata	cowpea, mung bean, moth bean
	Holotrichia sp.	wide range of plants
Mung bean	*Aphis craccivora*	many legumenous plants
Moth bean	*Amsacta moorei*	wide range of vegetation
	Bemisia tabaci	wide range of crop & fruits
	Callosobruchus sp.	pulses and legume tree seed
	Chrotogonus sp	grasses, millets
	Empoasca kerri	-do-
	Epilachna sp.	*Kharif* crops
	Heliothis armigera	wide range of crops
	Herse convolvuli	—
	Mylabris sp.	pulses, cucurbits
	Nezara viridula	pulse crops, fodder plants
	Spodoptera spp.	wide range of plants
Pearlmillet	*Amsacta moorei*	wide range of vegetation
	Bagrada sp.	pulses, brassica
	Chrotogonus sp.	grasses
	Cylindrothorax sp.	
	Holotrichia spp.	wide range of plants
	Myllocerus spp.	grasses, tree seedlings
	Nezara viridula	grasses, pulses
	Rhinyptia sp.	—

REFERENCES

Altieri, M.A., D.K. Letourneau, and J.R.Davis, 1983. Developing sustainable agroecosystems. *Bioscience* 33(1) : 45-49.

Chandra, H. 1987. Web of spider (*Stegpduphus sarasinorum* Karsch) a good insect trap. *Plant Protection Bulletin* (Ind.) 39(3) : 38-39.

Dempster, J.P. and T.H. Coaker, 1974. Diversification of crop ecosystems as a means of controlling pests. In (Jones D.P. and Solomon M.E. eds) *Biology in pest and disease control*. John willey. pp. 106-114.

Kareiva, P. 1983. Influence of vegetation texture on herbivour populations : resource concentration and herbievore movement. In (Renno R. ed.) *Variable Plants and Herbivores in Natural and managed systems.* Academic Press.

Rathore, M.P.S. 1995. Insect Pests in Agroforestry. Working Paper No. 70. ICRAF.

Risch, S.J., D. Andow, and M.A. Altieri, 1983. Agroecosystem diversity and Pest Control : Data tentative conclusions and new research directions. *Environmental Entomology* 12(3) : 625-629.

Saxena K.N. and A. Basit, 1982. Interferance with the establishment of the leafhopper *Amrasca devastars* on its host plants by certain non host plants. In Proc. V Intl. Symp. Insect Plant Relationship. Wageningen, Netherlands.

17

Diseases of Desert Trees and their Management

R. Raj Bhansali and S.K. Jindal

The trees grown under desert conditions need greater attention of the specialists for the development of integrated pest management (IMP) scheme for boosting forest coverage. The technology for the improvement, production and protection of multipurpose trees such as khejri (*Prosopis cineraria*), rohida (*Tecomella undulata*), kumat (*Acacia senegal*), Israeli babool (*A. tortilis*), babul (*A. nilotica*), jharberi (*Ziziphus nummularia*), neem (*Azadirachta indica*), mitha jal pillu (*Salvadora oleoides*), khara jal (*S. persica*), kair (*Capparis decidua*), phog (*Calligonum polygonoides*) etc., needs special emphasis of various disciplines for large scale plantation of superior fast growing saplings (Fig.10). Some of these trees are highly remunerative and useful for the livestock because of their highly nutritive and palatable leaves. The leaves are rich in crude protein (10 to 15 per cent) and calcium (1.5 to 3.4 per cent), and low in crude fibre (10 to 20 per cent).

Diseases of arid zone trees are among the major constraints to obtain healthy production of trees per unit area, time and inputs. Therefore, disease management is of prime importance to minimise the mortality of saplings at nursery stage and in the field at adult stage. In this regard, logically an efficient disease management programme should be most effective in accurate diagnosis of disease inciting agents. Three important facets of the disease management i.e. identification of pathogen, systematic approach for diagnosis of the disease or disease complex and application of plant protection measures, are highly significant for formulating host as well as location specific technology. As regard the control of plant diseases, chemical application alone is highly visible and effective method, but has become most controversial due to environmental hazards and danger for life supporting system of the desert fauna and flora. The modern techniques of disease management are based on accurate knowledge of pests and diseases; understanding of economic, environmental, cultural, genetic and microbiological factors, and judicious use of pesticides (Tattar, 1978 and Fry, 1982). The various interacting factors necessary for the disease development have been illustrated by the plant disease triangle (Fig.11).

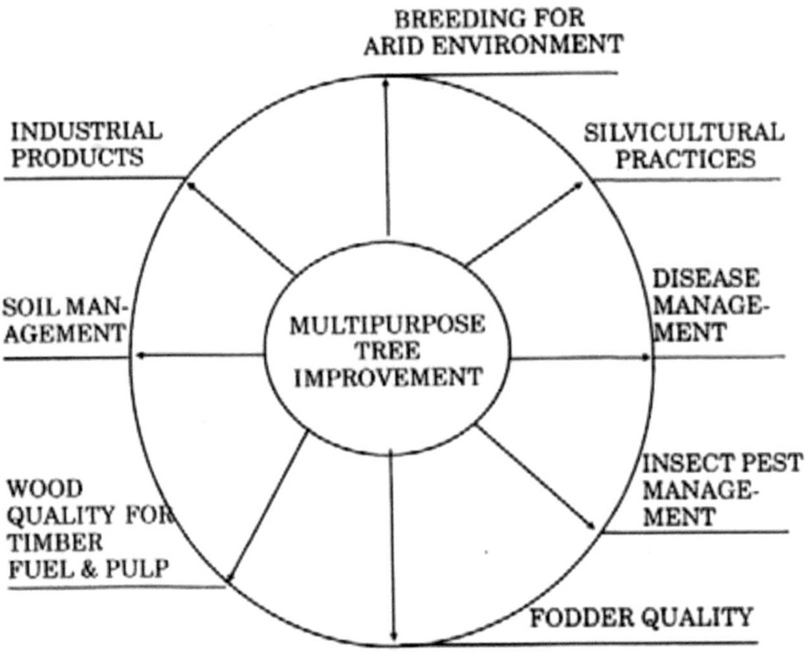

Figure 10. Various disciplines involved in multipurpose tree improvement programme

Though disease is a complex phenomenon which occurs due to the interaction between the host (tree), parasite (causing agent) and environment. Any change in arid and semi-arid environment in response to precipitation, light intensity, temperature and other cultural and plant production practices such as manuring, weeding, hoeing, spacing between the plants etc., affects the disease development under desert ecosystem. Similarly disease resistant/tolerant genotypes (host) also help in reducing the disease occurrence. The virulence of the pathogen can also be altered by

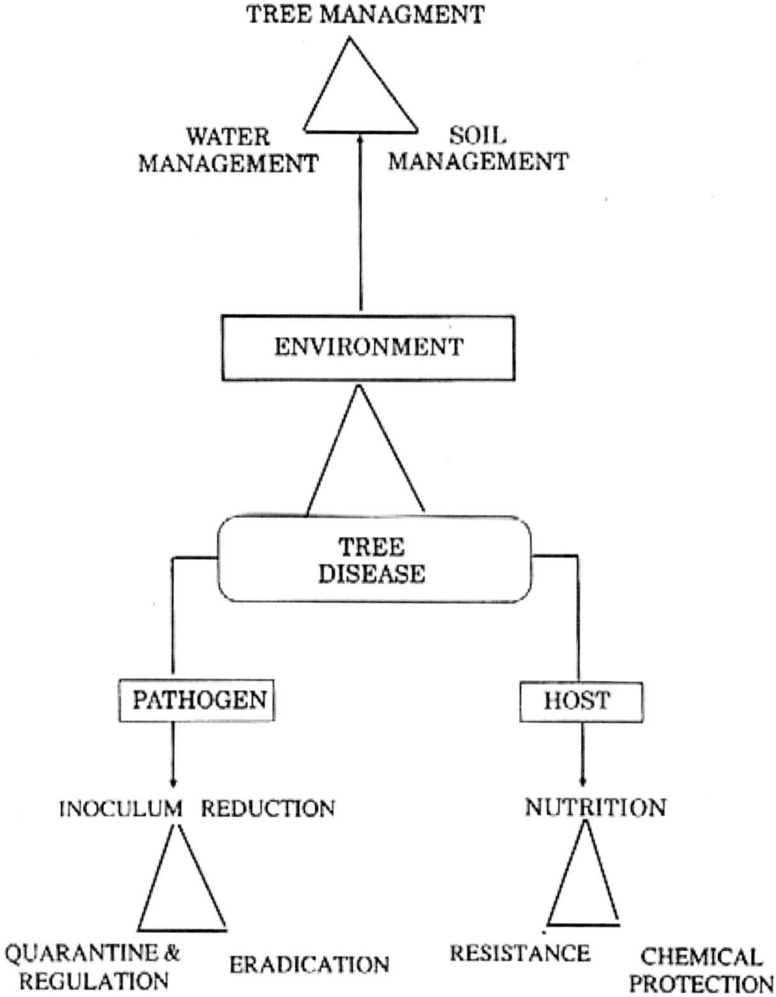

Figure 11. Factors involved in plant disease complex

manipulation of the cropping system and environmental factors. Any control measures which has objective to suppress the pathogen is helpful in reducing severity of plant disease. There are numerous ways by which this may be accomplished as depicted in Fig. 12.

In this chapter the important diseases of desert tree species with special reference to their management in nursery is described.

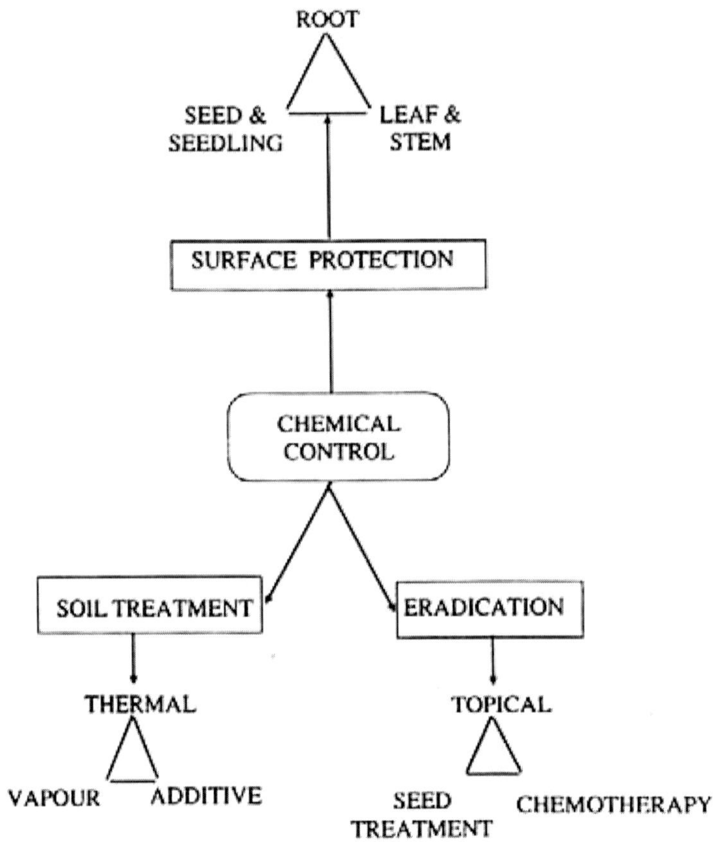

Figure 12. Principles of plant disease control with chemicals in arid zone trees

NURSERY DISEASES

Seeds of the forest trees are major source for raising saplings in nursery. Like those of agricultural and horticultural crops, seeds obtained from forest trees carry a wide variety of pathogenic microorganisms like bacteria, fungi and even viruses (Foster, 1956 and Bakshi 1976). The microorganisms penetrate into the seed coat, and even to the embryo while seeds are still on the tree or get in during the storage. The old seeds and poor storing conditions contaminate the seeds with various saprophytic fungi and bacteria which causes the rotting of seeds, poor germination and unhealthy seedlings. Many times seedling mortality also occur in nursery beds due to soil borne and water borne diseases, known as damping-off.

Damping-off

Found all over the world, damping-off diseases are common to most of forest seedlings, have wide range of hosts and causes serious damage to nursery plants (Tattar, 1978). The pathogens involved belong to the genus *Rhizoctonia*, *Fusarium*, *Sclerotium*, *Pythium*, *Phytophthora* and *Cylindrocladium*. In the Indian desert ecosystem, mortality of the seedlings at nursery stage is commonly observed in *A. nilotica*, *T. undulata*, *C. decidua* and *A. indica*.

Mortality in the seedlings occurs usually at two stages viz., pre-emergence and post-emergence. Pre-emergence mortality (before germination of seeds) occurs in heavily contaminated seeds and soils. The post-emergence mortality occurs either soon after the germination or also root-rot in the older seedlings. This causes poor growth of seedlings and sometimes even complete death. Sometimes sparse emergence of seeds may wrongly be interpreted as poor viability of the seeds. Under post-emergence mortality, the seedlings are attacked on hypocotyl and/or on roots. The pathogen spreads rapidly in the young tissues of seedlings causing wilt type symptoms due to the rotting at the seedling base or above the soft part. The poor soil moisture and drainage aeration also enhances the mortality of seedlings. The symptoms of damping-off develop more commonly during summers or due to sudden increase in temperature and in heavy soils.

Management of disease may be accomplished by cultural practices, chemical control or combination of both.

Chemical control of damping-off is achieved by seed treatment(s), application of fungicides to soil and soil fumigation. dithiocarbamates (thiram), carbendazim (bavistin), quintogene (brassicol), mancozeb (dithane M-45) are commonly used for seed treatment in desert trees. Thiram, a non-systemic dithiocarbamate fungicide also serves as a bird repellent. This chemical is more effective against pre-emergence than post-emergence mortality (Vaartaja, 1964). There are several new systemic and non-systemic seed dresser fungicides currently being used on agronomic crops but need to be tested for their efficacy. The doses of chemical may be fixed on type of host, pathogen(s) and method of raising seedlings in nursery, but usually it varies from 0.1 to 0.3 per cent active ingredient.

In nursery seed beds where losses due to damping-off are very severe or where other control measures are not successful, soil fumigation often is used to eliminate the diseases. Soil fumigation is accomplished with formalin and methyl bromide (Gibson; 1956 Seymour and Cordell, 1979). These fumigants are highly phytotoxic and therefore, should be applied to the nursery beds atleast 10 to 15 days before sowing. The toxic effects of the chemicals are removed through volatilization and can be checked by smelling the treated soil (Smith and Bega, 1966). In nursery beds, formaldehyde (1 per cent) can be applied by mixing with soil/potting mixture. Commercial grade fomaldehyde, (38 to 40 per cent 250 ml diluted in 10 litres of water) may be applied as soil drench per m^2 of nursery bed. A thick polythene sheet should be spread over the treated bed with edges of the sheet buried in the soil so that vapours released from the fumigants gradually diffuse into the soil. The polythene sheet is removed after 48 hrs of treatment. The fumigant also kills the undesirable weed seeds, nematodes and soil borne insects. As fumigants are highly poisonous, proper care should be taken in their use. To manage damping-off disease, an integrated approach involving measures such as sanitation, cultural practices, healthy seeds and use of formaldehyde should be adopted where infested soils are unavoidable.

Seed Borne Fungi

Tree seed mycoflora belongs to diverse groups of fungi. Many species are usually considered as un-important due to weak pathogens but under favourable conditions may cause considerable damage. Studies conducted on seeds of various arid zone tree species have shown association of various mycoflora responsible for poor germination and seed rotting. Maximum infection of seed borne fungi was recorded in seeds collected from ground than directly from the canopy. The mycoflora associated with seeds mainly belong to the species *Aspergillus, Penicillium, Mucor, Rhizopus, Fusarium, Cladosporium, Pythium* and *Colletotrichum*; and of bacterial origin. The fungal infections on seeds are mostly in group and rarely in isolated form. The growth of seedlings was severely affected when fungal contamination load on seeds is heavy. Bacterial infection causes a complete rotting of seed/ cotyledons/ embryos. The common range of hosts for seed borne fungi are: *A. indica, C. decidua, Simmondsia chinensis, Acacia* spp., *Prosopis* spp. and *T. undulata*. In case of jojoba (*S. chinensis*), seeds are usually associated with fungal microflora species of *Alternaria, Curvularia, Phoma, Botryodiplodia* and *Phomopsis*.

Damage of seeds during storage and mortality of seedlings in nursery pose serious problems in forest trees. These problems can be managed effectively through integrated control such as quarantine, seed quality certification, proper storage conditions and chemical treatment. Since most of the arid zone forest species are propagated by seeds, the quarantine measures by screening unhealthy (infected) seeds will be advantageous for checking movement of infected material from one place to another so that spread of diseases can be checked. This will help in maintaining seed quality according to ISTA standards and for raising quality forest in arid zone. For the quality seeds, seed should be processed in following ways:

(i) Discarding infected/mycoflora associated seed during collection operations.

(ii) Proper cleaning and extraction of seed.

(iii) Drying of seed to the moisture content of 6.5 to 12.0 per cent.

(iv) Storage under dry conditions and at a temperature which limits fungal growth.

Proper precautions at harvesting time and during post harvest operations are necessary to reduce infection and damage.

Seed treatment is useful in reducing losses due to fungal deterioration, raising disease-free stock and to avoid further pathogen spread. Seed treatment is most convenient and less expensive method of disease control. Seeds may be treated mechanically, physically, chemically or biologically. Surface sterilization of seed prevents attack on developing seedlings due to seed borne pathogens. The simple method such as treatment with hot water, hot air, solar radiation and aerated steam is also very helpful in reducing the fungal/bacterial inoculum on seeds. Application of fungicide protects the seed by inhibiting or killing seed borne pathogens and also helps in reducing seed decay and seedling disease. Seeds may be treated chemically as dry seed, seed dip, pelleting, fumigation, slurry and mist treatment. In case of severe infection, dipping and pelleting of seed are most effective in controlling a wide range of seed borne fungi.

Various systemic (carbendazim and quintogene) and non-systemic fungicides (mancozeb, captafol, captan, thiram and mercurial fungicides) are effective for the control of seed borne diseases. Biological control of seed borne fungi has also been attempted by using micro-organisms such as *Trichoderma*, *Penicillium*, *Bacillus* and plant extracts of *A. indica*, *Annona squamosa* and *Calotropis procera*.

STEM DISEASES

The most common and damaging stem diseases in the arid zone forest trees are : canker, anthracnose, sooty mold, die-back and heart rot. These diseases occur on *A. nilotica*, *P. cineraria*, *A. indica*, *Albizzia lebbek*, *T. undulata*, *Z. nummularia* and *Salvadora* spp (Raj Bhasali, 1990; Dwivedi, 1993). Die-back disease occurs in young seedlings as well as in adult trees. Various microorganisms (*Fusarium* spp, *Curvularia, Colletotrichum, Botryodiplodia*) are responsible for this disease. Fungal infections generally take place through open wounds, cracks, broken branches etc. These fungi are soil borne, air borne and sometime seed borne in nature. The diseased plants may be treated by spraying copper based fungicides and dithiocarbamates (0.2 per cent).

During monsoon seasons even a weak parasite such as *Torula herbarum* may cause twig mold to young *Tecomella* trees. The affected trees usually give die-back symptoms as complete loss of foliage occurs under severe conditions (Fig.13). The disease is more pronounced after rainy season.

Figure 13. Die-back symptoms on young *Tecomella* tree

The heart rot (wood rotting) fungus attacks *Acacia* spp., *Prosopis* spp. and *Albizzia* spp. The wood rotting fungi *Fomes, Phillinus* and *Inonotus* belong to family polyporaceae. The fruiting bodies (sporophores) usually appear on the tree trunk after rains

Figure 14. Fruiting bodies of *Phillinus* on *Prosopis cineraria*

are over (Fig. 14). Initially a knob or small outgrowth attached to the host tree appears on the wound of stem. These bodies gradually develop into large structures by attaining annual growth which indicates perennial nature of the fungus. The sporophores are sessile, hard, woody and easily detachable from the host. The sporophore body consists of four major regions i.e. pileus, context tube layers, pores and basidiospores. Disease can be minimised by avoiding injury to trees, dressing of wounds by copper based fungicides and illicit lopping. The desert trees are generally pruned/lopped to get fodder, fuel and timber, therefore, timely dressing of the cut ends is essential to minimise the attack of micro-organisms. Commercial wound paints based on copper fungicides should be applied over the surface of both pruned and bark wounds to prevent internal decay and also to act as an aid in wound healing. The fruiting bodies should be collected and destroyed by burning.

FOLIAR DISEASES

The desert forest trees growing in arid and semi-arid areas are susceptible to variety of foliar diseases. These diseases are sporadic and cause variable damage depending on environmental conditions and genetic make up of the tree species. Most desert trees are deciduous in nature and therefore loss due to foliar pathogens has limited scope, but severe attack sometimes leads to complete defoliation producing die back syndrome and even to the complete death of tree. Foliar diseases are commonly attributed to variety of fungal and bacterial pathogens. Removal of infected seedlings, trees or tree parts should form a standard cultural practice to reduce the

Figure 15. Leaf spots on *Eucalyptus* seedling in nursery

amount of inoculum. In the nursery where attack of foliage pathogens is frequent, seedlings density in beds should be reduced as it favours the development of foliage diseases (Fig. 15). Chemical control measures could also be applied either as prophylatic measures throughout the disease hazard period or only following disease initiation to arrest further spread and damage. The wide spectrum fungicide such as copper based fungicide, dithiocarbamates, difolatan, carbendazim, methylthiophonate, and mancozeb can also be used. Witches broom, little leaf, phyllody and viral leaf spots are also foliar diseases which occur in *Salvadora, Zizyphus, Eucalyptus* and *A. indica* species.

ROOT ROTS

Root rot diseases cause severe losses to a variety of desert tree species. These rots are mostly caused by *Ganoderma* sp. which is saprophyte or weak parasite (Fig. 16). The disease is reported from all over India especilly from the Punjab, Rajasthan, Haryana, Uttar Pradesh, Madhya Pradesh and Andhra Pradesh, mostly on leguminous trees, though it also occurs on other tree species such

Figure 16. Fruiting bodies of *Ganoderma* on *Acacia tortilis*

as *A. indica* and *S. oleoides* (Bakshi, 1971; Raj Bhansali, 1990). Infected trees exhibit symptoms initially with the pale foliage from top crown to downwards. The disease spreads gradually in plant and blocks vascular system causing defoliation and drying of the trees. The fungus mostly survive on dead and likely to dead root tissues. Though the disease is soil borne in nature but active growth of mycelium does not occure in soil. Disease spreads among the trees through root to root contact. The desert conditions favour the disease development. The young fruiting bodies appear on the tree near the ground portion as white buttons of mycelium on severely attacked trees. Sporophores are initially stalked and corky, later on structures become woody and upper surfaces are shiny and zoned (Fig. 17). Older bodies turn brown to reddish brown on drying and having minute pores. The disease can be minimised by through proper preparation of plantation site, through pulverization and exposing the soil to solar radiation. The old stems, residual roots and rootlets should be removed and burnt before raising the new plantation. Construction of isolation trenches (0.3 m wide, 0.7 m deep and 1.5 m long) across the line also prevents the entry of pathogen (Bakshi, 1976). Non-host tree species like *Ailanthus*, *Bombax* or *P. juliflora* are suggested to be raised in between the susceptible hosts to reduce the risks of *Ganoderma* attack.

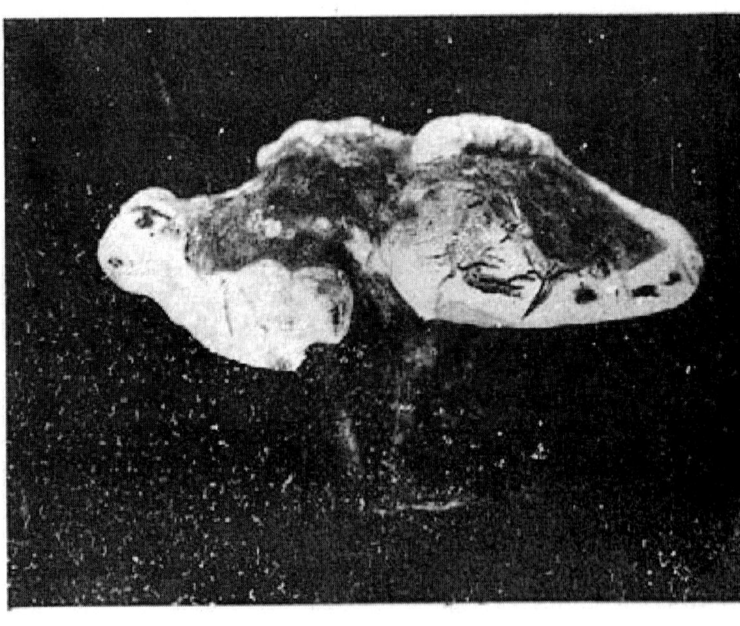

Figure 17 Sporophore of *Ganoderma*

Effective disease management in forest trees specially in nursery plants requires that nurserymen should use an integrated disease management approach consisting of coordinated combination of preventive, cultural, biological and chemical procedures.

REFERENCES

Bakshi, B.K. 1971. Indian polyporaceae on Trees and Timber. Indian Council Agric. Res. New Delhi.

Bakshi, B.K. 1976. Forest Pathology - Principles and Practice in Forestry. FRI Pub. Dehra Dun pp. 400.

Dwivedi, A.P. 1993. Babul (*Acacia nilotica*) - a multipurpose tree of dry areas. Sci. Pub. Jodhpur 131-161.

Foster, A.A. 1956. Diseases of the forest nurseries in Georgia. Plant Dis. Rep. 50 : 69-70.

Fry, W.E. 1982. Principles of plant disease management, Ist Ed. Academic Press, New York.

Gibson, I.A.S. 1956. Sowing density and damping-off in pine seedlings. E. African Agric. J. 21 : 183-188.

Raj Bhansali, R. 1990. Occurrence of wood rotting fungi on forest trees of arid zone. Annual Report, CAZRI, Jodhpur pp 74.

Seymour, C.P. and C.E. Cordell 1979. Control of charocal root rot with methyl bromide in forest nurseries. Southern J. Appl. For. 3 : 104-108.

Smith, R.S., Jr. and R.V. Bega 1966. Root disease control by fumigation in forest nurseries. Plant Dis. Rep. 50 : 245-248.

Tattar, T. A. 1978. Diseases of Shade Trees. Acad. Press. New York.

Vaartaja, O., 1964. Chemical treatment of seed beds to control nursery diseases. Bot. Rev. 30 : 1-92.

18

Important Fodder Trees of Arid Zone and their Management

J.C. Tewari

The developing world is facing an acute fodder shortage as serious as that of petrolium crisis. The production of green fodder on cultivated land is handicapped because of insufficient cultivated fodder lands, lack of irrigation facilities and low economic returns from fodder crops as compared to cash crops.

In certain parts of Western Rajasthan, probably more animals feed on shrubs and trees than on grasses and pasture legumes. The leaf fodder of some trees is almost as nutritions as that of leguminous fodder crops. From some of the trees, the leaf fodder production can be comparable with grass production from pastures. Besides trees have the potential to grow under harsh climatic conditions and on the lands where arable crops can not grow successfully. Fortunately, for many parts of arid region, fast growing varieties of multipurpose trees are available. This paper deals with important fodder tree species and their management practices in arid and semi arid regions of India.

IMPORTANT FODDER SPECIES AND THEIR MANAGEMENT

Although a number of tree species is used as top feed in arid and semi arid tracts of India, studies at Central Arid Zone Research Institute, Jodhpur revealed that *Prosopis cineraria* (Khejri), *Zizyphus nummularia* (Bordi), *Hardwickia binata* (Anjan) and *Colophospermum mopane* (Mopane) are most promising. Following discussion summerizes the distribution, characteristics, management practices etc. in respect of these four tree species.

Prosopis cineraria

It occurs in the arid and semi-arid regions of Rajasthan, Gujrat, South-east Panjab, parts of Maharashtra and dry tracts of south India. It also occurs in Sind and Baluchistan in Pakistan. Naturally, the species is distributed in arid climate with rainfall ranging between 150 to 500 mm with the summer temperatures varying between 45 to 48°C. It can grow on a variety of soils but deep sandy loam soil with adequate availability of moisture in the lower layers is optimum. It commonly grows on the agricultural fields. There is a popular saying that "Death will not visit a man even at the time of famine if he has a Khejri plant, a goat and a camel".

Generally it is lopped heavily during winter season. In double cropped area, in the month of October, whereas in single cropped area the lopping is done from November to December. Bhimaya et al. (1963) standedrised the lopping intensities. A mature tree (about 7 m height and 83 cm DBH) when fully lopped yielded around 59 kg leaf fodder, whereas 66 and 33 percent lopping could yield 28.5 and 20 kg tree^{-1}, respectively. Lopping at an interval of four years, produced 238 per cent higher leaf fodder than the annual lopping (Table 35). Gupta and Mathur (1974) reported maximum leaf fodder production when trees were lopped in winter season. The nutritive value of this leaf fodder was also of very high order (Table 36).

TABLE 35. Fresh weight of leaves (kg Tree^{-1})of *Prosopis cineraria* (Average of 8 years)

Treatments	Diameter class (cm)			
	20-25	25-30	30-35	30-40
Recurrent lopping year after year	6.3	6.8	7.7	10.4
Lopping after four years	21.3	22.6	26.6	27.2
% increase	238.0	232.8	241.5	161.5

Source : Bhimaya et al. (1963)

TABLE 36. Percentage proximate components of *P. cineraria* leaves

Component	% of dry matter
Crude protein	11.9
Ether extract	2.9
Crude fibre	17.5
Nitrogen free extract	43.5
Ash	8.1
Phosphorus	0.38
Calcium	2.10

Source : Gupta and Mathur (1974)

The paltability of the leaves is very good and it improves as the tennin content decreases with the maturity of leaves. The percentage digestibility of leaves lopped in June-July and September-October were found to be 41 and 48 per cent for dry matter, 39 and 56 per cent for crude protein, 10 and 17 per cent for ether extract, and 69 and 72 per cent for nitrogen free extract, respectively. The total digestible nutrients in leaves lopped in September-October were higher than those lopped in June-July. In addition to the leaf fodder, green pods are also used as vegetable and riped pods as animal feed. From a grown up tree (Height 668.7 cm; dbh 29.9 cm) nearly 2.8 kg pod can be obtained annually (Kaul et al., 1964).

Harsh et al., (1987), observed that in arid regions the leaf production is proportionate to hight and crown diameter, whereas collar diameter and DBH has no affect on the leaf production. Annual lopping was found to decrease the leaf fodder production.

Ziziphus nummularia

Ziziphus sp. are the important source of fodder in arid and semi arid regions. Z. nummularia leaves provide a sustenance when the ground cover is depleted. It can grow on almost all types of habitats. On flat alluvial plains, its natural regeneration is encouraged by the farmers in their own field as well as in 'oran lands' (common lands). It has been reported to be confined to Persia (Iran), Baluchistan (Pakistan) and India. In India, large stands of this species are found in Punjab, Haryana, Rajasthan, Gujrat, Uttar Pradesh and Madhya Pradesh. In Western Rajasthan, the highest density of (720 plants ha^{-1}) was recorded in Nagaur followed by Jodhpur (471 plants ha^{-1}), Churu (405 plants ha^{-1}), Bikaner (390 plants ha^{-1}) and Jaisalmer (65 plants ha^{-1}) (Shankar, 1981). The leaf fodder production estimated by Shankar (1981), Muthana (1981) and Kaul and Ganguli (1983) was maximum from burried pediment (169 kg ha^{-1}) and lowest from pediment and lower pediments (1.12 kg ha^{-1}).

The studies made by Ganguli *et al.*, (1964) on the nutritive value of 'Pala' (leaves of *Z. nummularia*) revealed higher concentration of crude protein, crude fiber, calcium and Phosporus in green leaves than in dry leaves. Average values of these components are given in (Table 37). Patel *et al.*, (1978) studied seasonal variations in the chemical composition of Pala leaves and observed that the leavels of nitrogen free extract (NFE) and ether extract (EE) increase from monsoon to summer, while crude fibre follows the reverse order during the same period (Table 38).

TABLE 37. Chemical composition (% on DM basis) of *pala*

Component	Percentage
Crude protein	11.70
Crude fibres	16.0
Calcium	1.60
Phosphorus	0.20

Source : Ganguli et al. (1964)

TABLE 38. Seasonal variation in the proximate components (% ON DM basis) of pala

Constituent/Season	Monsoon	Winter	Summer
Crude proteins	14.2	14.6	14.8
Crude fibres	20.2	11.7	8.2
Ether extract	3.2	5.2	5.9
Nitrogen free extract	51.8	57.8	60.7
Calcium	1.70	2.17	2.07
Phosphorus	0.19	0.16	0.16

Source : Patel et al. (1978)

Recent studies on improvement of *Z. nummularia* undertaken at CAZRI revealed that budding of root stock of *Z. nummularia* with improved varieties of 'Ber' i.e. 'Seb' and 'Gola' not-only improve the fruit size and yield but also increases leaf fodder production with better quality.

Hardwickia binata

Native of Tamil Nadu, Karnataka, Andhra Pradesh, Maharashtra, Madhya Pradesh and Uttar Pradesh, introduced in 1962 at CAZRI, Jodhpur. It performs well in arid part and has a great potential as fodder tree. In its natural habitat it grows gregariously in areas adjoining Maharashtra and Karnataka while in the remaining areas it grows mixed with other miscellaneous broad leaf species. It is a tree of dry hot climate with scanty to moderate rainfall, intense heat during summer and a long period of drought. It can grow in rainfall ranging between 250 mm to 1000 mm. It grows on a wide variety of habitats.Under natural habitat the tree attain a height of 20 to 25 m in 50 to 60 years, while under favourable condition it can attain a height up to 30 m and girth up to 2 m, though on poor and isolated sites, the length of clear bole is short. The tree pollards well and coppices indifferently. The tree reproduces from root suckers also (Rao, 1953). It responds to lopping very favourably.

In arid region, the disposed seeds germinate well during the monsoon period, but establishment is very very poor because of erratic and long dry spell. Therefore, plantation is done by planting out entire seedling, raised in the nursery during Feb.–March. The nursery raised seedlings were planted on the field during the onset of monsoon. It was observed that it grows very fast upto the age of 6 yrs and can attain a height of 3 m in 3 yrs and 4.5 m in 6 yrs. The Annual increment in first 6 years was around 27.5 cm/tree/year, while 6th year and onwards it was 87.0 cm/tree/yr.

At the age of 7th year it was lopped upto 50 and 75 per cent of its total canopy and leaf fodder production was recorded. At 75 and 50 per cent lopping it produced 6.7 kg and 4.5 kg green leaves, respectively. The same trees were again lopped in subsequent year and was observed that leaf fodder production declined sharply. At 75 per cent looping it was only 315 g tree^{-1}, while at 50 per cent it was 175 g tree^{-1}. Though the plant earlier responded to lopping very well but due to severe drought (60 mm rains from June to Dec. in 1987) the leaf fodder production was very poor. Therefore, in a severe drought year this plant should not be lopped at all (Harsh et al., 1987). It was also observed that in the unlopped trees the height increment was only 13 to 20 cm tree^{-1}, however, in case of 75 per cent lopping, the annual increment was 35 to 55 cm tree^{-1}.

The leaves of *H. binata* are categorized as good fodder (Krishnaswami and Reddy, 1955; Behari and Prasad, 1968) containing 8.96 per cent crude protein, 4.05 per cent ether extract, 2.28 per cent calcium and 0.13 per cent phosphorus on dry matter basis (Prasad et al., 1974). The average digestibility co-efficients for dry matter, crude protein, ether extract crude fibre, and N-free extract were 47.4, 20.4, 49.0 and 56.9, respectively. The total digestible nutrients were 47.69 per cent.

Colophospermum mopane

It is an exotic tree introduced from south central Africa (Zimbabwe). It has great fodder potential for arid parts of India. In favourable circumstances the 'mopane' attains the height of 12 m and diameter upto 120 cm, but in poor soil and harsh climatic conditions, the height attained is less than 8 m. At CAZRI, Jodhpur it attained a height of 3.25 m in 9 yrs. In the natural habitat 'mopane' is not more than a small shrub as the basaltic changes contract and expand with drought or rain, napping the roots and thus restricting the natural growth.

The leaves are valuable fodder, as they can retain their food value for weeks even after shading from tree together. The pods are flat, thine, kidney shaped and contain single seed. Mopane produces valuable amount of seeds varying from few to thousands depending upon the locality and climate. The seeds are about 1.25 cm in diameter, flat kidney-shaped and highly resinous. The seeds are unpalatable to almost all animals. The seeds spread by wind.

At CAZRI, Jodhpur, it was introduced in 1963 from Zimbabwe. In 10 years it attained an height of 439 cm with collar diameter of 15.2 cm, and DBH about 7.6 cm. The observations revealed that mean annual increment in height of mopane during first six years of growth was 50 cm tree^{-1}. Lopping of the trees in eight year upto 50 and 75 per cent of its canopy produced 3.3 and 6.7 kg tree^{-1} green fodder, respectively. The heavy intensity of lopping had no effect on growth of mopane trees.

CONCLUSIONS

In addition to native fodder tree species like *P. cineraria* and *Z.nummularia*, some exotic species could be used for fodder production in arid and semi arid areas. However, to ascertain the exact fodder potential of these introduced species, it is essential to conduct studies on the growth behaviour, propagation, lopping regime etc. on long term basis. The introduction of mixture of native and exotic tree species in term of lands i.e., waste lands, common lands, pastures etc. may be very effective for top feed production.

REFERENCES

Behari, M and Prasad R. 1988. Fodder potential of trees. *Allahabad Farmer* 42 (4) : 245-248.

Bhimaya, C.P., R.N. Kaul and B.N. Ganguli 1963. Studies on lopping intensities of *Prosopis cineraria*. *Indian For*. 90 : 19-23.

Ganguli, B.W. R.N. Kaul, and K.T.N. Nambiar 1964. Preliminary studies on a few top-feed species. *Ann. Arid Zone* 3 : 31-37.

Gupta, M.L.and C.S. Mathur 1974. Studies on seasonal variation in the chemical composition of Khejri (*Prosopis spicigera* Linn). *Indian For.* 100 : 269-73.

Harsh, L.N., J.C. Tewari, D.S. Patwal and G.L. Meena 1987. Prediction of leaf fodder of Prosopis cineraria using different growth parameters *Annual Report*, CAZRI, Jodhpur.

Kaul, R.N. and B.N. Ganguli 1963. Fodder potential of *Ziziphus* in the scrub grazing lands of arid zone. *Indian For.* 89 : 623-630.

Kaul, R.N. R.P. Goswami and B.K. Chitnis 1964. Growth attributes for predicting pod and seed yield of *Prosopis specigera*. *Sci. and Cult.* 30 : 682-685.

Krishnaswami, V.S. and T. Reddy 1955. *A short report on the fodder, milk supply and land utilization problems of the state of Madras.* Supdt. Govt. Press, Madras.

Muthana, K.D. 1981. Silvicultural aspects of *Ziziphus nummularia. (Bordi) a shrub of the Indian Arid Zone. Its Role in Silvipasture* (Mann H.S. and S.K. Saxena eds.). CAZRI, Monograph No. 13 pp. 15-18.

Patel, B.M. P.C. Shukla, V.V. Mistry and P.S. Patel 1978. Survey of animal nutrition in Rajasthan. *Proc. Indian National Sci. Acad.* 44 : 282-295.

Prasad, D.A., K.J. Reddy and C.S. Subba Reddy 1974. Chemical composition and nutritive value of Anjan (*Hardwickia binata*) leaves. *Indian J. Animal Sci.* 44 : 150 -152.

Rao, H.S. 1953. Vegetative propagation and forest tree Improvement. *Indian For.* 79 : 176-183.

Shankar, V. 1981. Distribution pattern of Jharber and its leaf fodder and bushwood production on different habitat and land use types. *Bordi (Ziziphus nummularia) A shrub of the Indian Arid Zone. Its role in silvipasture* (H.S. Mann and S.K. Saxena Eds.). CAZRI, Bulletin No. 13 pp 10-14.

19

Genetic Improvement of Tree Species Suitable for Arid Zone

K.R. Solanki

Maximization of production per unit area to narrow the gap between demand and supply of the forest products is possible only by artificial plantation (man-made forests). Any scheme on afforestation, however, in addition to proper management, requires correct choice of species, i.e. seeds of genetically improved populations and modern silvicultural treament. It is in this context that tree breeding assumes importance. There is increasing realization among the foresters that Genetics and Tree Breeding has highest potential among various forestry practices for increasing forest production per unit area.

IMPORTANCE OF SEED

The quality of seed used for afforestation is of utmost economic importance. Forest trees are extensively heterogeneous in quality and seeding ability. More prolific seed bearers are often inferior genotypes. The silvicultural practice of thinning and eliminating the undesirable types results in a limited improvement. A breeder, however, aims at production of types having superior heritable nature i.e. superior genotypes.

The application of genetics and breeding principles vis-a-vis certified seed has already resulted in green revolution in agriculture. With the general acceptance that plantation is a better alternative for regeneration, there is a realization to exploit hereditary principles to tailor trees to produce maximum biomass. Although there has been a definite trend towards greater use of phenotypically superior trees as seed source, the recourse to the use of genetically superior seed in majority of forestry operation is still in its infancy.

Use of certified seed in afforestation programme has remained neglected, though it has greater importance in forestry than agriculture because of the longer life span of tree and the error of having used poor quality seed can not be rectified. Recently it has been realized that superior quality seed can repay many times its initial cost. This has resulted in the collection of seed now only from seed stands, plus trees, or from specially established plantation or seed orchards.

CHOICE OF SPECIES

The fundamentals of planning a forestry programme in arid zones involve lot of thinking in respect of selection of species. In arid zone because of harsh climate, limited availability of moisture and inhibitation in soils, the choice is limited when large areas are proposed to be planted. Care should be taken that the species selected have multipurpose utility and have consideration of energy, social and economic aspect and special emphasis on the locality aspect. Plantation of nitrogen fixing leguminous trees which have faster growth and production,ameliorate wastelands,provide nutritional forages and serve as renewable energy source in the highly degraded lands and environmental set up, would be advantageous.

At Central Arid Zone Research Institute, Jodhpur tree species identified for different habitats and soil types of desert zone are: *Acacia tortilis, Prosopis cineraria, Prosopis juliflora, Albizzia lebbek, Azadirachta indica, Acacia senegal, Tecomella undulata, Dichrostachys glomerata, Colophospermum mopane, Brasilettia mollis, Hardwickia binata, Tamarix articulata, Cassia siamea, Eucalyptus camaldulensis, E. hybrid.*

For sand stone rocky sites : *Acacia tortilis, Prosopis juliflora, Albizzia lebbek, Butea monosperma, Acacia senegal,Commiphora wightii, Cassia auriculata, Anogeissus pendula, Bogwellia serrata.*

For saline area : *Salvadora oleoides, S. persica, Tamarix articulata, prosopis juliflora, P. tamarugo.*

For shifting sand dunes : *Acacia tortilis, Prosopis juliflora, Prosopis cineraria, Acacia senegal, Dichrosatchys glomerata,Colophospermum mopane, Albizzia lebbek, Tamarix articulata, Calligonum polygonoides.*

Preferred indigenous tree species in the arid areas are : *Prosopis cineraria, Acacia senegal, Ziziphus* sp. *Capparis* *decidua*, *Tecomella undulata* and *Salvadora oleoides*. These tree species, However, are extremely slow growing. *Prosopis juliflora* for fast biomass production, *Colophospermum mopane* and *Dichrostachys glomerata* for fodder purpose, and *E. camaldulensis* are few exotics suited for low rainfall area. Indigenous tree species suited to better rain fall areas of arid zone are *Azadirachta indica* and *Albizzia lebbek*. Under irrigated conditions, planting of *Dalbergia sissoo*, *Acacia nilotica*, *A. indica* and *Leucaena leucocephala* has been found useful.

EXPLOITATION OF AVAILABLE GENETIC VARIABILITY

In multipurpose trees, simultaneous improvement in more than one character is needed. In general, trees with better establishment and faster growth are desirable. The information available from other tree improvement programme can not be applied as such to arid zone trees because the plant density is generally very low and pollinators and mating systems are likely to be different. Also there are chances of genetic drift and beside genotype, change is an important factor for establishment and survival of trees in arid zone. Still to begin with the basic steps are same i.e. understanding the patterns and extent of variations, *in situ* collection of variability, partitioning of variability into genotypic and environmental components, identification of superior genotypes and establishment of seed orchards. Salient achievements in tree improvement programme are discussed in the following paragraphs.

Prosopis cineraria

Among the indigenous tree species, *Prosopis cineraria* (Khejri) is the most important multipurpose tree species of this region. This tree is normally slow growing and because of this lacuna it is generally not included in the afforestation programmes. In order to understand the extent and pattern of variability and its exploitation, detailed surveys were undertaken to collect the promising germplasm, and progeny trials and seed orchards of the superior genotypes were established.

Natural Variability

In western Rajasthan, natural stands had a tree height ranging from 4.6 to 20.4 m, forking height 1.2 to 4.6 m, diameter at breast height (DBH) 0.14 to 1.5 m, collar dia. 0.16 to 1.6 m, branches per tree 2 to 9 and canopy dia. 1.2 to 9.1 m. Coefficient of variation in these traits ranged from 26 to 64 per cent. Large variations were also observed in pod length, spiny habit and leaflet size. Better phenotypes were available in the districts of Barmer and Sikar (Kackar et al., 1936a). All the morphological traits were significantly associated with each other. However, phenotypic traits relating to seed are more reliable than others as these are neither affected by the age of individual trees nor associated with other morphological traits.

Maximum variability for pod length (8.6 to 26.0 cm), number of seeds per pod (2.0 to 17.0) and 100-seed weight (4.0 to 7.4 g) was observed for Jodhpur, Nagaur and Churu provenance, respectively. The range of variation was higher for arid regions than for semi-arid. For example, in Barmer provenance, CV, was 30 per cent for all the three characters (Kackar et al., 1986b). Wide variability was observed in crude protein, reducing sugars and total sugars contents in pods even from the same provenance. The range was 8.05 to 15.05 per cent for crude protein, 0.83 to 4.67 per cent for reducing sugars and 6.04 to 15.83 per cent for total sugars. In juvenile progenies, peroxidase isoenzymes showed large variations at some loci, both within and between the progenies.

Germplasm Collection

Collection of seeds of plus trees and establishing their single plant progenies, was the major approach for collection of germplasm. Seeds collected during June were sown in the same month and two months old seedlings were transplanted during August in the same year. The establishment was more than 90 percent by this method (Kackar et al., 1991). Using this approach more than 3000 trees representing single plant progenies of 142 plus trees have been established. Vegetative approach for collection and conservation of germplasm was also employed and 200 plants were established using air layering technique. *In vitro* collection methods are gaining importance and can be suitable for arid zone trees. Initial success in culturing nodal segments in field on ms medium containing Bavistin and Streptocycline has been achieved (Solanki et al., 1991a).

Progeny trials : These trials were initiated in 1984 with a view to study the variability within and between the progenies for estimating the genetic parameters as a prerequisite for conceiving selection criteria. Sufficient variability has since been observed, both within and between the progenies which have aided in making selection of superior genotypes. Progeny from accession No. 264 (from Jodhpur) has shown consistently better growth performance over the years (Table 39).

TABLE 39. Mean tree height of half sib progenies trial established during 1984

Provenance Number	Height (m)						
	1986	1987	1988	1989	1990	1991	1992
304 Nagaur	1.39	1.69	2.06	2.36	2.98	3.21	3.34
316 Churu	0.87	0.87	0.98	1.00	1.28	1.45	1.45
249 Jaisalmer	0.77	0.81	0.76	0.81	1.01	1.27	1.28
216 Jalore	1.12	1.23	1.44	1.69	2.07	2.34	2.37
264 Jodhpur	1.55	1.91	2.29	2.62	3.37	3.77	3.91
206 Barmer	1.21	1.44	1.74	2.16	2.81	2.97	3.16
274 Tonk	1.19	1.51	1.72	2.11	2.71	2.97	3.04
284 Jaipur	0.91	0.94	1.10	1.28	1.63	1.80	1.84
301 Sikar	1.08	1.03	1.22	1.46	1.91	2.10	2.14
297 Jhujhunu	0.88	0.84	0.83	0.87	1.09	1.25	1.26
321 Bikaner	0.92	1.01	1.22	1.43	1.89	2.10	2.12
Mean	1.08	1.21	1.40	1.62	2.07	2.29	2.36
CD (P = 0.05)	0.21	0.29	0.37	0.46	0.57	1.45	1.37

Seed orchards : The performance of some elite selections of *P. cineraria* in terms of height growth was even better than *A. tortilis* which is very fast grwoing exotic species. *P. cineraria* attained a height of 7.3 m against 6.5 m of stand of *A. tortilis* in 6 years (Solanki et al., 1984a). Such identified elite trees were used to establish the seed orchards. In the seed orchard, there are 156 elite trees (mean height in 1992 being 5.30 m in 8 years). The best tree attained a height of 8.60 m. During 1991, 80 trees having relatively poor growth in the seed orchard were lopped. The average performance of even these trees was very encouraging, mean fresh biomass at first lopping was 10.93 kg (the best being 32 kg).

Clonal Propogation of Superior Genotypes

Prosopis cineraria exhibits wide variability in natural stands in morphological, biochemical and seed related traits. During surveys, a few trees showed exceptionally good growth and other adaptive traits. These plus trees are the result of natural selection. In a genetic improvement programme, it is desirable to bring together such genotypes spread over a large area for breeding purpose and establishment of clonal seed orchards.

Seeds collected from plus trees from natural stands gave heterogeneous population as expected. Since conventional approaches to obtain homogeneous populations take long time, propogation of superior trees was followed. The propogation technique through air layering standardized in 1983-84 was followed. During rainy season (July-August), branches of 10 to 15 mm dia. were selected, girdles 15 mm apart were made to remove the bark and rooting hormone Seradex B_3 was applied before covering it with a layer of clay. The entire band was then covered with polyethylene sheet and tied firmly at both the ends to check the moisture loss. Root initiation was perceptible after about a month. After full development of the roots, the branches were cut and planted as saplings in pots (Solanki et al., 1984b). This technique has been useful to contribute to about 200 plants now standing in the clonal seed orchard (Solanki et al., 1991a). Good rainfall ensures a high success of layering and their establishment (Solanki et al., 1991b).

Tecomella undulata

Tecomella undulata (Rohida) belongs to family Bignoniaceae which is known for many woody perennials with large size attractive coloured flowers. Seeds of *T. undulata* are very light, papery and easily get dispersed by wind. Several other plants of this family viz. *Tecoma stans*, a large yellow flowered shrub; *T. jasmonoides*, a pink flowered climber; *T.capensis, T.radicans* and *T. grandiflora* with red/orange flowers; are the attraction of many Indian gardens. *T. undulata*, however, is the only indigenous species (Brandis, 1972). In Indian arid zone, it is locally called as "Marwar Teak" or "Desert Teak" because of its wood quality. Presently it is one of the most important desert timber tree species. A large scale export of its carved furniture during last three decades has resulted in a great demand for its wood. Consequently the population of this species, especially large, bold and healthy plants has decreased drastically to a stage that it is now listed as threatened species (Pandey et al., 1983, Saxena, 1993).

In western Rajasthan, *Rohida* (*T. undulata*) is generally recorded on sandy undulating hummocky and undulating interdunal plains with deep to very deep soils. Barmer and Sikar districts are two distinct concentration zones of high density of Rohida. In Barmer District, Chohtan, Barmer and north Bhakasar tehsiles show high density of *Rohida* (20 to 60 pts ha^{-1}) on flat and hummocky interdunal areas. Although this is a low rainfall tract (150 to 250 mm), the soil moisture in lower depth is present round the year which possibly sustains high tree density. *Rohida* is generally associated with *khejri* (*P. cineraria*) community in these areas. Low dunes and longitudinal ridges of these three tehsils are also occupied by this species with 10 to 20 plants ha^{-1}

Natural variability

The mean, range and standard deviation for various quantitative characters studied by (Solanki et al., 1993) are reported in Table 40.

TABLE 40. Quantitative characters of *T. undulata*

Characters	Range	Mean
Tree height (m)	12.8-20	5.78
Forking height (m)	0-4.2	1.63
Branches per tree	1-15	—
Basal diameter (cm)	7.1-136.8	—
DBH (cm)	3.8-130.4	—

Maximum coefficient of variability was observed for basal diameter (61.3 per cent) followed by DBH (59.2 per cent) and minimum for leaf size 1.19 to 6.08. Peak flowering period is from end of February to middle of March. The flowering duration varies from 59 to 103 days for individual tree and upto 135 days for the whole population. It has been observed that almost all the flower buds appearing in the months of November, December and January fell before or after opening and only a few flowers developed in the month of February set fruits. Fruit setting varied from 0.64 per cent for selfing to 3.94 per cent for cross pollination indicating there by the presence of self compatability. Apomixis, an important mechanism for fixing heterotic advantage, was found absent. (Jindal et al., 1985a). A wide variability was observed in 198 trees studied. Maximum coefficient of variability for tree height, DBH, basal diameter and canopy diameter were observed in Barmer district. (Jindal et al., 1987). Jindal et al., (1985b) evaluated 185 genotypes of *T. undulata* in natural habitats for various quantitative characteristics and a wide variation in height, basal diameter and DBH. Maximum frequency and abundance was observed in interdunal areas of Barmer district.

Progeny trials : Jindal et al. (1991) studied the height of ten randomly selected plants of 24 individual tree progenies during first four months. Heritability and genetic advancement showed a decreasing trend with increase in age. Correlation of juvenile height at different stages with mean height of one year old progenies in field was non-significant suggesting that selection for height and juvenile stage in nursery may not be effective.

Tree height of eleven half sib plant progenies established in 1984 was recorded for six growth years. Progenies showing good growth in the initial stages were not necessarily performing the same in subsequent years. There were significant differences among the progenies during all the six growth years. Estimates of family heritability, single tree heritability and genetic gains from family and mass selection indicated that the gains were low with both the methods. As components of variance tend to stabilize after fourth year of growth, selection made during early stages may not be rewarding. Low values of heritability and genetic gains from family and single tree selection suggested that there is need for other selection procedure for higher genetic gains for the tree height (Jindal et al., 1992).

In the progeny trial comprising of eleven half sib plant progenies established at CAZRI, Jodhpur during 1984 revealed that beside segregation of height, diameter at breast height (DBH) etc. there was segregation for flower colour in ten progenies, suggesting the outcrossing nature of this species. The observations showed that birds like purple sun bird (*Nectarinia asiatica*) visit the flowers for nectar and may be acting as pollinators.

Acacias

A. senegal is an important fuel and gum yielding tree found on rocky habitats. In an exploratory survey undertaken during 1984, maximum variability was found in Barmer district. In 52 trees enough variability was observed in various pod characters like pod length (4.2 to 10.3 cm), seeds/pod (2.8 to 7.0) and 100 seed weight (7.40 to 18.88 g). Variation in different traits was also observed in progenies of 13 plus trees established in 1985 at Jodhpur, Bhopalgarh and Jaisalmer. At present there are about 700 trees in the field representing 18 single plant progenies.

In *A. tortilis*, an exotic fast growing multipurpose tree species, twelve progenies were evaluated under arid (Jodhpur) and Semi-arid (Pali) conditions. Components of variance for height tend to stabilize by third year (Solanki et al., 1992). Provenance trials of *A. tortilis* ub. sp. *raddiana* and *A. albida* have also been established during 1988-89.

In *A. nilotica* collections were made from eleven provenance in Gujarat, and a replicated trial was established during 1990. *A. nilotica* ub. sp. *cupressiformis* is found in Pali district and its concentration is very high around Sadri. During survey, variation in height, DBH and crown compactness was observed. Seed from eleven plus trees has been used to establish progeny trial of this sub species during 1991. Earlier studies on single plant progeny indicate that this sub species can perform well under Jodhpur conditions and there is segregation with respect to compactness of crown.

CONCLUSIONS

From the forgoing discussion it is revealed that trees are important constituent of arid zone ecosystem and provide/satisfy multifarius needs of desert farmers. Research on genetic improvement clearly indicated that variability for almost all the silvicultural characters is available in the Thar Desert. Further, there is a possibility of selection of plus or elite trees keeping in view of the objectives and needs of particular zone. However, considering the long life span and extreme climatic conditions proper planning and testing of the material through multilocation trials are required so as realize to the genetic potential to maximum.

REFERENCES

Brandis, D. 1972. The forest flora of North West and Central India. Reprinted by BS MP Singh, pp.352-353.

Jindal, S.K. K.R., Solanki and N.L. Kackar 1985a. Phenology and breeding systems of Rohida (*Tecomella undulata* (Sm.) (Seem). Indian J. For. 8 : 317-320.

Jindal, S.K., N.L. Kackar and K.R. Solanki 1985b. Evaluation of *Tecomella undulata* (sm.) Seem germplasm from western Rajasthan *in situ*. Trans Isdt & Ucds. 10 (2) : 33- 37.

Jindal, S.K., N.L. Kackar and K.R. Solanki 1987. Germplasm collection and genetic vriability in Rohida (*Tecomella undulata*) (sm.) Seem in western Rajasthan. Indian J. For. 10 : 52-55.

Jindal, S.K., Manjit Singh, K.R. Solanki and N.L. Kackar 1991. Variability and changes in genetic parameters of height in juvenile progenies of *Tecomella undulata* (Sm.) Seem. J. Tree Sci. 10 : 25-28.

Jindal, S.K., Manjit Singh, K.R. Solanki and N.L. Kackar 1992. Changes in genetic parameters and ranks of tree height over six growth years in *Tecomella undulata* (Sm) Seem. Silvae Genetica. 41 : 213-216.

Kackar, N.L., K.R. Solanki and S.K. Jindal 1986a. Variation for morphological characters of *Prosopis cineraria* in natural stands. My Forest 22 (2) : 129-134.

Kackar, N.L., K.R. Solanki and S.K. Jindal, 1986b. Variation in fruit and seed characters of *Prosopis cineraria* (L.) Mac Bride in the Thar desert. Indian J. of For. 9 : 113-115.

Kackar, N.L., K.R. Solanki and Manjit Singh 1991. Genetic improvement and other related studies in *Prosopis cineraria* a review. Symposium on Khejri, Kalarsar Vikas Samiti, Jaipur, March 8-9, 1991.

Pandey, R.P., B.V. Shetty and S.K. Malhotra 1983. A preliminary census of rare and threatened plants of Rajasthan In (S.K. Jain and R.R. Rao Eds.). An assessment of threatened plants of India, Botanical Survey of India. Botanic Garden, Hawrah, 1983 pp. 52-62.

Saxena, V.S. 1993. Propagating non-traditional species in plantations workshop. Proc. Tree Improvement and Provenance Research, Department of Forest, Govt. of Rajasthan, Jaipur : 135-145.

Solanki, K.R., K.D. Muthana, R.S. Paroda and N.L. Kackar 1984a. Performance of the elite trees of *Prosopis cineraria* in comparison to *Acacia tortilis* in western Rajasthan. Forage Res. 10 : 111-113.

Solanki, K.R., N.L. Kackar and S.K. Jindal 1984b. Propagation in *Prosopis cineraria* (L.) Mac Bride by air layering. Curr. Sci. 53 : 1166-1167.

Solanki, K.R., N.L. Kackar and S.K. Jindal 1986. Air layering in *Prosopis cineraria* (L.) mac Bride. Indian For. 112 : 202-207.

Solanki, K.R., Manjit Singh, S.K. Jindal and N.L. Kackar 1991a. Approaches for collection of germplasm of multipurpose trees of Thar desert. Golden Jubilee Symposium, Indian Soc. Gene. and Pl. Breed., New Delhi, Feb. 12-15, 1991.

Solanki, K.R. and N.L. Kackar 1991b. Studies on micropropagation and vegetative propagation in *Prosopis cineraria*. Workshop on vegetative propogation of the biotechnology of tree improvement, Tirupati, March 23-26 1991 pp. 153-159.

Solanki, K.R., N.L. Kackar, Manjit Singh and S.K. Jindal 1992. *Prosopis cineraria* - Collection of germplasm its evaluation and exploitation. Intnl. Symp. on Prosopis spp. University of Durham, England. 25-31 July pp 157-173.

Solanki, K.R., Manjit Singh, S.K. Jindal and N.L. Kackar 1992. Annual changes in components of variance of height in *Acacia tortilis* over five year growth period. J. Tree Sci. 11: 21-24.

Solanki, K.R., N.L. Kackar, S.K. Jindal and Manjit Singh 1993. Assessment and Conservation of variability in arid zone tree species. Proc. Workshop on Tree Improvement and Provenance Research, Forest Department, Rajasthan. pp. 7-24.

20

Improved Varieties of Pasture Grasses and their Cultivation

M.S. Yadav and M.P. Rajora

The importance of grasses for rangelands are three fold, viz. livestock feed, soil builders and as an aid in soil conservation. Grasses are not only the cheapest source of animal feed in the tropics but also the highest potential yielders of starch and protein than any other crop. Rangelands in western Rajasthan occur on sandy, rocky stretches, saline and other areas. However, eighty per cent of the total grazing lands in western Rajasthan are poor having an estimated forage production of less than 300 kg ha^{-1}. Therefore, improvement of rangelands by reseeding with high yielding varieties of grasses is imperative. Based on the edaphic characters, following grass species have been identified.

Sand Dunes and Sandy Plains : *Lasiurus sindicus, P. turgidum, Panicum antidotale* and *Cymbopogon jwarancusa.*

Well drained sandy alluvial soil: *C. ciliaris* and *C. setigerus.* Sandy clay loam to clay soil : *Dichanthium annulatum* predominants on such types of soils in high rainfall zones (360 mm and above). Hilly and piedmont regions : *Sehima nervosum*; *Dichanthium annulatum.* Low lying heavy saline soils : *Sporobolus marginatus, S. coromandelianus* and *Chloris virgata.*

Grasses species for different agroclimatic regions are as follows:

	Climatic zones/habitat	Rainfall (mm)	Grass species
1.	Arid zone sandy soil	125-375	*Lasiurus sindicus, Cenchrus ciliaris, Panicum antidotale, Panicum turgidum*
2.	Semi-arid	375-825	*Cenchrus ciliaris*
3.	Humid arid and sub lumid	1000-2000	*Setaria sphacclata, Panicum maximum, B. decumbens*
	(a)Well drained soil (b) Marshy areas		*Pennisetum purpureum, Brachiaria brizantha, B. mutica.*
4.	Saline and alkaline soils		*Chloris gayna, Paspalum notatum, Andropogon elongatum, Cynodon* sp. *Sporobolus* sp.
5.	Lower hills (upto 4000')		*Chrysopogon fulvus, Themeda anathera, Arundinella* sp. *Dichanthium annulatum, Pennisetum orientale*
6.	Upper hills (4000 to 10000')		*Dactylus glomerata, Bromis inermis , Poa pratense, Lolium perenire , Phallium pratense.*

CHARACTERISTICS OF HIGH YIELDING VARIETIES

At Central Arid Zone Research Institute, Jodhpur a broad based germplasm of over 1600 collections from diverse indigenous and exotic sources has been assembled. This genetic wealth was subjected to critical screening and comprehensive evaluation tests with a view to isolate desirable variants for direct utilization in pasture. An active variety reconstruction programme employing modern methods of pasture improvement on *Lasiurus sindicus* and *Cenchrus ciliaris* and *C. setigerus* was taken up and a number of high yielding varieties of promising pasture species were identified. In order to assess their potential value as forage grass species for drought affected areas, the selected varieties were also studied for their seed viability, dormancy, germination inhibitors, root systems, reaction to defoliation, crude protein content at different stages of harvest, palatability, dry matter yield and seed yield.

The desirable features of these widely adapted selections *inter alia* include, high dry matter yield, greater nutritive and seeding value (greater leafiness, better palatability, digestibility and intake), greater persistency, aggressiveness, compatability with associate species, quick regeneration ability for multicut, and reasonable seed production capability. The improved varieties of promising grass species with their important characteristics are discussed in the following paragraphs.

Lasirus sindicus (*Sewan*) : It is one of the promising grasses of desert which can survive under extreme arid condition and remain productive for a period of 5 to 10 years. The recommended strains are CAZRI 817, CAZRI 319, CAZRI 351, CAZRI M-30-5 and CAZRI 295 producing 6000 to 8500 kg ha^{-1} of dry nutritious forage 6-13 per cent protein depending upon the precipitation and establishment. This grass naturally grows in extensive areas in forest of Rajasthan desert and is responsible for maintaining desert breeds of cattle namely "Tharparkar" and Red sindhi ", Seed rate is 4 to 6 kg ha^{-1}.

Cenchrus ciliaris (*Dhaman* and *Anjan*) : Dhaman may stand supreme for rehabilitation of arid waste, the stabilization of sand dune or provision of nutritious forage for sheep and cattle. Its importance as a pasture grass for range seeding has been recognized almost throughout the arid and semi-arid regions of the world. It is a tufted perennial grass with a stout root stock. The recommended varieties are CAZRI-358, IGFRI-3108, Biloela and Molopo giving a range of productivity from 4000 to 5500 kg ha^{-1} when seeded at the rate of 3.5 to 4.5 kg ha^{-1}. Three to four cuts can be obtained from August to April. It contain 8 to12 per cent crude protein. Dhaman grass produces 75 kg ha^{-1} seed and the best time to harvest for hay is at the appearance of flower heads.

Cenchrus setigerus (*Moda Dhaman*) : It is one of the major grass of arid zone pasture. It is perennial forming clumps from somewhat bulbous base. The plant with a height of 50 to 70 cm with an average number of 25 tillers per clump. There are various growth forms i.e. prostrate, tufty nature of growth and less ciliated seeds. It prefers better moisture condition. The recommended varieties are CAZRI-76, CAZRI 296, CAZRI 175, CAZRI 415 and pusa yellow Anjan having potential to give 3000 to 4500 kg ha^{-1} of dry matter. Crude protein content of grass varies from 5 to 11 per cent of the dry matter for different cuttings. It gives 2-3 cuts in a year from August to April. It produces 100 kg ha^{-1} seed.

Panicum antidotale (*Gramna*) : It is also a perennial hardygrass, which is stemmy but cut frequently or grazed closely with short intervals. Productivity varies from 2000 to 3200 kgha^{-1} dry matter. The recommended varieties are CAZRI 331, CAZRI 333, CAZRI 379 and CAZRI 347. Gramna is often found on sand dunes and is an excellent sand binder. It is a deep rooted drought resistant grass. It can reach a height of 240 cm. The lower half of stems have large nodes with hard internodes having a bamboo, like appearance. For dry areas, the recommended seed rate is 2.5 kg ha^{-1}.

Dichanthium annulatum (*Karad*) : In arid zone of Rajasthan and Marvel, Palwan and Zincoo in other parts of the country. It occurs in depressions and basin like situations of the low rainfall areas of the arid zone. Karad is a perennial tufted grass with creeping rhizomatous stem. The plant grows to an height of 75 cm at maturity producing about 103 tillers per plant. Diploid, tetraploid and haxaploid species of this grass occur in nature. The diploids are tropical; hexaploids are of south African type whereas tetraploids may occur both in tropical as well as mediterranean regions.CAZRI 490, CAZRI 491, IGFRI-495 and marvel-1, are the recommended varieties of Karad. Dry matter yield of these varieties ranges between 3500 to 5500 kg ha^{-1} (2-3 cuttings). The recommended seed rate is 2kg ha^{-1}. The seed being very light, it is necessary to mix them with soil before sowing. The crude protein content of grass at flowering stage varies from 3.87 to 6.97 per cent.

21

Social Issues in Agroforestry Land Use Systems on Arid Lands

L.P. Bharara

Agroforestry land use systems hold serious social and economic implications in the fragile ecosystem of arid areas. Ecological deterioration and accentuation of natural hazards affecting balanced land use system has been occurring in the region due to human and livestock pressure. Fallow periods are shortened and cultivation is extended to the areas unsuitable for cultivation. Communities with static farming and traditional technology face various problems of over-exploitation of natural resources in order to meet the increasing demands of food, fodder,fuel and water resources. This has resulted in low productivity, low income, poverty, deforestation and nomadic and semi-nomadic pattern of living associated with migration and pastoralism. Continuance of high fertility norms encouraging high growth rate of population, massive illiteracy, traditionalism and socio- religious beliefs, traditional agricultural and animal husbandry practices retarding adoption of new technology, accelerate the impact of deterioration of land use system. This causes misuse of land use system involving cultivation of sub-marginal lands, overgrazing of rangelands, over cultivation, deforestation and cutting of vegetation and depletion of wood cover affecting agroforestry land use system in the area.

Systematic attempts to identify appropriate agroforestry systems for different localities and improve them for higher productivity and better economic returns had seldom been made until the early 1970's. Research on agroforestry and silvipastoral systems were initiated at the Central Arid Zone Research Institute (CAZRI) Jodhpur, Inida in the early seventies. Various other studies (Malhotra, 1981, 1984; Shankarnarayan, 1984; Shankarnarayan et al. 1987; GOI, 1987; Baumer, 1991; Ballabh & Saxena, 1991 & 1992; Bharara, 1970, 1988, 1989, 1991, 1992; Jodha, 1982, 1985, 1986; UNEP, 1991 & 1993) highlight on agroforestry practices, need importance but lack vital social issues.

However, not much research has been done on the socio-economic aspects, issues and prospects of agroforestry land use systems on arid lands. Empirical studies on farm trees, their significance, changes over time, relationship with agraian structure and farming systems, etc. are still lacking. Even well known Indian village surveys done in the post-independence period do not make a mention of village trees at all (Ballabh and Saxena, 1992). The present paper discusses socio-economic aspects, issues and prospects of agroforestry land use systems in dryland agriculture specifically arid zone of Rajasthan.

CONCEPT AND IMPORTANCE

Agroforestry has a long tradition in India. The socio-religious fabric of the people is inter-woven to a very great extent with, raising, caring for and respecting the trees which are integrated extensively in the crop and livestock production systems under different agro-climatic and other local conditions. The multitier crop combinations in the area meet cash and household requirements. The *Khejri* (*Prosopis cineraria*) and crop combination in the hot arid region meet needs for fodder, small timber and food.

Arable farming has dominated dryland agriculture so far. It is only recently that alternate land use systems are receiving attention in dryland agriculture. This is because these systems give stability to the resource-poor, small and marginal farmers. The cereal grains and pulse crops grown by the dryland farmers are just enough to meet his family requirements. The fruit products invariably results in marketable surplus and thus bring ready cash to the dryland farmers.

Fodder is an important requirement of dryland farmer. The pinch of fodder shortage is felt more during the drought years. Some of the fruit trees, apart from yielding fruits, also yield valuable leaves. These leaves form nutritious fodder for cattle. e.g. *ber* leaves contain about 10 per cent crude protein and are sold at high price, especially during drought years. Apart from leaves the fruit trees also yield wood from the annual prunings, which can be used as fuelwood.

Fruit trees provide employment for the farmer and his family members during off-season also. Agroforestry system generates additional employment opportunities for rural people. Nursery raising, budding, pruning, harvesting and packing will be a good source of employment for children, women and school drop-outs. It is estimated that by the turn of century, at least 25 per cent of the rural population has to be provided with alternate employment other than traditional farming. This is very important to check large scale migration of rural people to urban areas in search of employment. Hence, agroforestry and the associated industries like processing, packing, canning, etc. would be potential employment for rural people.

SOCIAL FACTORS INFLUENCING AGROFORESTRY

In fact agroforestry is the product of not only physical environment but also of social and technological frame work such as land ownership and tenancy system, size, shape and layout of plantation, pressure of population, irrigation systems, impliments, market, attitude of farmer and his literacy status, etc. Hence, launching a direct attack on technological problems alone will not be sufficient to ensure success for agroforestry development plans and package programmes, but simultaneous attention has to be given to the social institutions which at the moment are impediment to the rapid development of agrofroestry. So much intimate is the relationship between agriculture and character of social institutions and customs that it becomes virtually impossible to decide which is cause and which is effect.

The apparent increase of agroforestry also appears to be linked to the social changes that have occurred over the past thirty or forty years. Population increase, institutional changes affecting natural resource use, and the linkage to wider markets push for the intensification of production. However, the way in which agroforestry has come into its own is rather through a circuitious route. It appears to be a private response initially to the ecological degradation and/or disappearance of village common lands, the major traditional source of fuel and fodder. If sufficient quantities are no longer available from common lands then attention turns to tree resources on one's own land. Farmers also realise the direct benefits to crops and are encouraging *Khejri* on that account as well. In effect, farmers are reinterpreting and integrating trees into their rainfed farming to enhance their personal resource base and intensify the production of crops, fuel, fodder and timber. In the process they are also decreasing the risk of degrading their own land.

SOCIAL GROUPS INFLUENCING AGROFORESTRY

The role that the tree and shrubs of agroforestry systems can play in animal production in the area is itself influenced by the role animals play in society. There are various kinds of social groups concerned with animal production and these have implications for agroforestry. The groups indentified are :

The Pastoral group

These are groups (*Raikas, Sindhi, Muslims*) where animal rearing is the main occupation, or even the only one. This group includes nomads, semi-nomads and transhumants. They are traditionally in close association with their animals, constantly searching for pasture and not interested in change. The pastoral people as a whole are considered particularly well-adapted to the use of their environment.

The Agropastoral group

These people (*Rajput, Jats, Bishnois*, etc.) are mainly dependent on agriculture but some also practise animal husbandry. These groups are in favour of settlement. They get benefit from any misfortune of the nomadic people, for example, in periods of drought, they can acquire livestock at cheaper rates. The agropastoral group is increasing faster than the pastoral group, and in many places, encourage desertification by continuously pushing cultivation towards the more arid areas that are less and less suitable for sustainable agriculture. Many development projects are changing their economy to a mixed farming one.

The Mixed farming group

This group consists mainly of settled farmers (*Rajputs, Jats, Ghanchis, Bishnois,* etc.) who invest in more or less sophisticated forms of fodder production and management of herds and flocks. There are several agroforestry technologies that could help this group achieve relatively ensured return, for example crops for fodder or small fodder banks.

Crop farmers

This group (*Rajputs, Jats, Mali, Bishnoi,* etc.) is very similar to the previous one, consists of farmers who are deeply involved in crop production and have few animals. They get fodder from the agricultural by-products and it would be expected that agroforestry would concern them more in respect of the improvement of cropping practices than for animal production. Agroforestry will also increase the amounts of agricultural by- products, many of which are a source of fodder.

ASSOCIATIONS AND INTERDEPENDENCE

The basic outlines of agroforestry systems are a number of associations and interdependencies between trees, field crops, and animal husbandry that are also linked to fuel and timber concerns. Trees, crops and animals are a tripod which farmers have adapted to the fluctuations in their environment with trees serving as the stabilizing element for both crops and livestock. The spread and development of this system also allows for the intensification of production on a long term sustainable basis. The most important and wide spread tree, perhaps as a result of social selection, is *Prosopis cineraria* locally called *Khejri*.

The *Khajri* is managed and harvested like a crop, the activities and sequencing of which are neatly integrated with farm operations, labour availability, animal and human needs, and natural cycles. First, the random spacing of trees presents no difficulty for animal powered operations. Draught animals, usually bullocks and camels, are yoked singly and can plough straight tight furrows through a thick stand with ease. Secondly, bean like pods are produced during the hot pre-monsoon season when vegetables are scarce. The young tender beans are a supplementary and free, vegetable particularly for the poor. Third, harvest occurs after *kharif* is over. This creates an extra month to month and a half of employment and wage labour during slack season, particularly where a second crop is not taken. Fourth, the *Khajri* is harvested by lopping its leaf bearing branches. These are dried and the leaves beaten off. The leaves are harvested during November and December. Fifth, the woody loppings are used as a domestic fuel and for other purposes such as firing bring bricks. This frees animal dung from use as a fuel, allowing most to go into compost. Sixth, management practices dovetail with agro-metereological cycles for the two cropping season. This is significant if a winter *rabi* crop is taken. The removal of the canopy opens the area to warmth and solar radiation during the cool winter season when it is needed.

Most farmers do leave a few branches to allow trees to spread. In fact some farmers carefully shape their trees to make as full a canopy as possible, although this is highly variable. In any event an assured source of seed is associated with religious practices. Shrines to field God are found on most farmer's fields and are always shaded by a *Khejri* which is never lopped or cut. The associations and interdependencies found in this system are apparently long recognized by farmers. However, as a dominant practice it appears to be a relatively recent phenomenon, gaining greater recognition over the past thirty years.

A mature tree can bring several hundred rupees as firewood when the need for cash is immediate. Farmers practicing rainfed agriculture do report, however, that they now have more trees than earlier, and they want more. They also report that they actively encourage and protect *khejri* seedling that appear in their fields whereas earlier they ignored or uprooted them during field operations. The practice also appears more prevalent with those farmers who, for more intensive land use and to shorten fallows, erect boundry walls to protect their crops from grazing animals. In the process the tree seedlings are also protected from grazing that destroys or retards the reproduction of trees.

SOCIO-ECONOMIC ADVANTAGES

The advantages of agroforestry in dryland agriculture are as follows:

(i) It yields extended range of products viz; food, fodder, fuel, fibre, fruits, fertilizer, etc.

(ii) It helps in utilisation of off season percipitation. Most of the drylands are cropped for a single season but if a perennial tree component is there in the system it would make use of the same efficiently.

(iii) Agroforestry in dryland helps soil and water conservation. The perennial component of the agroforestry system either on contours or in some other geometry helps in reducing soil and water losses.

(iv) It imparts stability and results in risk reduction of the two or more components of the system, even if one fails, the other would give a harvest.

(v) Presence of trees in agroforestry system gives extended management options for the farmers, i.e. the trees can be harvested either for food or fuel or pole or small timber as per farmers requirements and market demands.

(vi) Agroforestry system checks the rise in soil temperature especially during summer months. Thus, it protects the soil micro-organism; which would be great benefit in crop production. Trees serve as shade for cattle.

(vii) It helps in recycling of nutrients from lower soil horizone. The deep rooted nature of most trees results in tapping of nutrients from deeper soil layers and returning them to plough layer through leaf drop and litter.

(viii) Off season family labour utilisation. During off season there would be no field work for family labour in dryland areas.

This time can, thus, be effectively used for harvesting pruning, marketing of the tree components in agroforestry systems.

(ix) Tree component of agro-forestry system results in improvement in the micro-climate of the whole system.

(x) Agroforestry results in improving and stabilizing the income of the farmers.

PERCEPTION TOWARDS AGROFORESTRY SYSTEMS

Of the various types of farming in practice, agriculture, pastoralism and agro-pastoralism are the most predominant. Agriculture as the main sustenance resource is followed by more than 60 per cent of the households of *Rajput*, *Peetal*, *Bishnoi* and *Rajpurohit* communities. While pastoralism is the main earning source of hardly 5 per cent of the total households belonging to *Raika* community. Agro-pastoralism is most commonly practiced by more than two-third of the total households of these communities having a combination of agriculture and livestock as the major sources of livelihood. Agriculture and pastoralism being more risk prone needs to be supplemented with other alternative farming types. The people preferred to have a multi-component system of farming types *viz.* agro-pastoral, agro-silvi-pastoral or agro-silvi-horticulture-cum pastoral (Table 41).

INDIGENOUS AGROFORESTRY AND ITS REVIVALS

A Biomass Production Oriented Strategy

The number of trees especially *khejri* and *Ber* bush colonies (called *malta*) within the plots and *matt* (shelterbelt) on the field borders, are different components of indigenous agroforestry systems in the villages of arid zone. They play important role in providing fodder, fuel and other materials besides keeping part of the crop land under natural vegetation. *khejri* tree are lopped for fodder and fuel every year after the end of *khejri* crop season. *Ber* bushes with deep roots are cut (harvested) before planting the crops. They again sprout and grow with the crops without competition. They are again cut after the harvest of field crops. The fodder from *khejri* called *loong* and from *ber* bush called *pala* are available every year. They are less affected by rain conditions compared to crop-based biomass. Hence they are also called desert farmer's natural insurance measures. During the drought years *loong* and *pala* fetch as high price as good grains. During good rain years the crop yields both grain and fodder around *khejri* trees and *ber* bush colonies are often higher than other parts of the same field.

TABLE 41. People's perception towards proportion of income and risks involved (%) in various agroforestry systems followed by the desert dwellers

Predominant agroforestry system	% Income from different components of agroforestry system and risks uncovered during average year				
	Agricu lture	Pastor alism	Silvicul ture	Horticul ture	Risk Uncovered
Agriculture	65	-	-	-	35
Pastoralism	-	60	-	-	40
Agro-pastoral ism	50	30	-	-	20
Pastoralism-cum-Agriculture	35	45	-	-	20
Agro-Silviculture	60	-	15	-	25
Agro-Silvi-pastoral	45	30	10	-	15
Agro-Horticul ture	55	-	-	15	30
Silvi-Pastoral	-	45	25	-	20
Agro-Pastoral -cum-Silvi-Horticulture	40	35	5	10	10

Trees and bushes have impact on micro-climate and they help in accumulation of humus, etc. Unlike trees and *ber* bush colonies within the field, the trees, bushes and shrubs constituting *matt* (Shelterbelt) are not harvested, except for collecting small material for fuel, thatching, etc. Thus, agro-forestry components perform serveral functions besides acting as renewable resources of fuel and fodder.

Complementary usage of Perennials and Annuals-Agroforestry

Provision of indigenous agroforestry clearly reflects this facet of farmer's strategy for higher and stable production of biomass. Perennials like *khejri* trees and *ber* bushes besides facilitating better performance of crop, act as more stable sources of biomass for the farmer. Table 42 presents bio-mass production from perennials and annuals including planted crops.

TABLE 42. Biomass yield during different rainfall years in an arid village of Jodhpur district.

Biomass yield	Drought year (1963-64)	Average year (1965-66)	Good year (1973-74)
Rainfall (mm)	159	259	473
Fodder yield (kgha^{-1})			
Top feeds	377	318	342
(Loong + Pala)	(78% pala)	(40% pala)	(53% pala)
Grass	85	250	1850
Crop by product	62	435	2160
Food Grain (kgha^{-1})	7	108	615
Fuel/fencing			
Thatching materials :			
Crop by-product (cart load)	0	2	4
Others (Cart load)	2	2	2

During the poor rainfall years top feeds (*loong* and *pala*) play important compensatory role in overall supplies of bio-mass (i.e.fodder). This is partly due to less sensitivity of these sources to fluctuations in rainfall and partly due to greater efforts by the farmers to harness them during the poor rainfall years. During good rainfall years by-products of crops contribute the largest share of total biomass production. But due to labour shortage and sense of complacency, biomass from agroforestry components is not fully harnessed. Not only *Pala* and *Loong* remain unharvested but in many cases *bajra* stalks are not harvested. There is also greater emphasis on collecting fuel, fencing and thatching materials (including from common property resources).

INCENTIVES

Quite encouraging signe are revealed by increase in number of *khejri* trees in crop fields; though, for other components e.g. *ber* bush colonies (*malta*) and shelterbelts (*matt*) situation is discouraging, as they are rapidly disappearing (Jodha, 1990). For promoting trees in the fields, as the current initiative of the farmer is a response to a need felt for drought period insurance, especially after *ber* bush in the fields declined following tractorisation.

Decline of CPRs (i.e. their area and productivity), also encouraged protection and growth of trees as a part of agroforestry, as an alternative source of biomass. This is partly responsible for privatisation of products of seasonal CPRs (i.e. top feed, fuel etc. from private lands traditionally used as CPRs during the off season). Hence, as CPRs decline, agroforestry may emerge as possible option to partly compensate the loss.

Tractor cultivation which discourages provision of trees or bushes in crop field and high cost and absence of institutional sanctions for protecting the exclusively private gains in the initial stage have little direct help to rural poor, modern scientific agroforestry is still unknown to the farmer.

There is little scope for effective public intervention, legal or otherwise, but some incentives to farmers might help. Discourage tractor cultivation; formal agroforestry programmes need to be sensitised to farmer's approaches to biomass production.

AGROFORESTRY POLICY FORMULATION

Agroforestry policy may be formulated to attain a specified goal, i.e. productivity per unit area over time, reduction in instability of production, socio-economic security of a family and rural community within a designated span of time.

The major criteria of agroforestry policy formulation may be in terms of people's particular needs, their goals and their resources. The policy should be politically acceptable, administratively feasible, economically sound, socially acceptable and capable of modifications as conditions change. The trees in each and every cultivated or uncultivated land with an individual may be marked on the map of the farm so that the indiscriminate felling may be stopped easily. If possible age of the trees or shrubs may also be entered in the pass book of every *khatedar* in rural areas.

The Consolidation of holdings, Ceiling Act and Land tenurial ownership may act in developing of the agroforestry. Some times these legislations become hinderance in development of agroforestry or vice-versa. Agroforestry can be developed by providing some incentive to the new allotees for raising trees alongwith crop cultivation. These trees after full development provide some remuneration or services to the person concerned even in scarcity period. Agroforestry would help in income stabilization over worst period and provide additional returns per unit area over space and time.

One of the most important policy implications relates to the perceptions of the various groups of people involved in agroforestry development in arid areas. On one hand, the farmer's themselves have clearly perceived the problems faced by them due to a decline in supply of needed forest products in the common forest land. They have started reacting to the situation by increasing the quantity of trees on their private farm land. They are very conscious and aware of their dependence on trees for the survival of their subsistence farming lifestyle. However, many government officials and policy planners perceive that the farmers lack an awareness of their dependence on trees and lacking knowledge, expertise, or ability regarding the cultivation and management of trees on their private land.

Traditional agroforestry system suggests that people take active interest in tree cultivation and protection provided it strengthens their production systems. The main policy implication for arid areas is, therefore, to develop trust between the intervening agency and the people. How to do ? This is a big question. This finding highlights the importance of research and action plan in the area of people's participation. A radical solution to this question might be that people would participate in agroforestry, tree protection, etc, provided the programme is economically beneficial to them. By corollary, they will not participate in agroforestry etc; for the sake of environment and for larger societal consideration. Therefore, the programme related to agroforestry should be based on economic considerations or enlarging their subsistence base. This is a necessary condition for success. Growing trees on private agricultural wastelands (as opposed to good quality land which can support more labour intensive annual crops) is both socially and economically desirable activity.

The function of agroforestry is precisely to look into problems of land use systems involving both agriculture and forestry and this would certainly need a change in our research approach. To begin with, there is a need to appreciate the fundamental characteristics inherent in all such systems : that they provide multiple outputs from the same unit of land in a sustainable way and that in some form or another interactions occur between the two or more constituent plant species, one of which at least is a woody perennial (tree or shrub, etc.). We are to deal with relatively elaborate systems for which programmes of investigations need to be established that encompass more complex space and time dimensions than those normally encountered in agricultrue and forestry. While we explore the problems and possibilities of intimately mixing trees/ agricultural crops and/or grasses sometimes in conjunction with animals as secondary producers, it is bound to bring into focus constraints requiring new lines of research approach like (i) shade tolerance of various agricultural species, (ii) forest species which protect the soils but do not reduce the energy levels on the understory, (iii) breeding programmes designed to obtain in both forest and tree species and in crops those attributes and characteristics necessary for successful and efficient intercropping, and (iv) the plant (topfeed) - animal interactions and their management.

A strategy for agroforestry animal husbandry based regeneration of arid areas should include :

(i) Strict operation of Rajasthan Soil and Water Conservation Act.

(ii) Land use according to its capability and need.

(iii) Stopping the allotment of denuded forest land for arable cultivation.

(iv) Prohibition of carriage of fuelwood and encouragement of community and farm forestry.

(v) Barring free and uncontrolled grazing in areas which are highly erodible and where erosion has already destroyed the fertile layer of the land.

(vi) Formulating a sound forest and vegetation policy.

(vii) Providing incentives for protection and plantation of trees in farm lands.

(viii) To enforce village panchayat social legislation for protection and plantation of trees in farm lands.

(ix) To encourage people's participation by enforcing vocational literacy programmes.

(x) Making use of traditional knowledge, attitude, aspirations, parception and motivation of the farmers influencing adoption of technology for increasing resource productivity.

(xi) To enforce desert care code highlighting socio- ecological ethics of environmental conservation strategies to be followed by the pastoralists during migration, at camping sites and while returning from migration. This includes :

(a) Establishment of pastoral net work viz; pastoralists, agro- pastoralists and grazing lands, coexistence, and a forum on grazing lands development.

(b) Manageable unit of livestock.

(c) Sustainable land use criteria for grazing of livestock.

(d) Judicious watering pattern to be followed.

(e) Maximum production with limited and diversified resources.

CONCLUSIONS

The future course of action for alleviating the misery of dryland farmers and boosting production and productivity from dryland should direct at the development of afforestation through agroforestry on watershed basis. Improvement and stability of agroforestry system in drylands is erroneous unless the Govt. move in a big way to create permanent assets with dryland farmers and provide a dependable infrastructure support for meeting input, credit and marketing needs. Voluntary agencies functioning in these areas can join hands in this exciting venture.

REFERENCES

Bharara, L.P. 1970. Social aspects of farm forestry in arid zone. Ann. Arid Zone, 9 : 36-44.

Bharara, L.P. 1988. Social policy for desertification in Rajasthan. In. Desertification Control, 7 (3) : 26-28.

Bharara, L.P. 1989. Socio-economic aspects of livestock migration and pastoral nomadism in drought-prone areas of Rajasthan. In. Livestock Economy of India. Oxford and I.B.H. Publishing Co., New Delhi, pp. 89-112.

Bharara, L.P. 1991. Socio-economic and legal aspects of grazing lands with emphasis on livestock migration and pastoralism in arid zone of Rajasthan, India. IV International Rangelands Congress. Montpellier, France, Serial 415.

Bharara, L.P. 1992. Population pressure and its impact on Arid Zone of Rajasthan. India World Congress on Natural Hazard Reduction, Vol. I. Institution of Engineers (India). New Delhi, pp.13-26.

Ballabh Vishwa and N.C. Saxena. 1992. Farm Forestry - A Review of issues and prospects. IRMA, Anand, India.

Ballabh Vishwa and N.C. Saxena. 1991. Socio-economic aspects of tree growing by farmers in south Asia. Institute of Rural Management,

Anand and Oxford Forestry Research Institute, Oxford.

Baumer, Michel. 1991. Animal production, Agroforestry and Similar Techniques, Agroforestry Abstracts, 4 (4) Nairobi, Kenya.

Govt. of India, 1987. Description and classification of wastelands. National Wastelands Development Board, Ministry of Environment & Forests.

Jodha, N.S. 1982. Decline of common property resources and its implication for livestock farming in arid region of western Rajasthan : Micro-level evidence. Paper Presented at the workshop on : Problems and prospects of livestock development, Institute of Development Studies, Jaipur.

Jodha, N.S. 1985. Population growth and the decline of common property resources in Rajasthan. India Population and Development Review, 2 : 247-264.

Jodha, N.S. 1986. Common property resources and rural poor in dry regions of India. Economic and Political Weekly. 21(27).

Malhotra, S.P. 1981. Traditional Agroforestry practices in arid zone of Rajasthan. Lecture delivered at the Summer Institute on Agroforestry in Arid and Semi-Arid Zones, CAZRI, Jodhpur.

Malhotra, S.P., H.S. Trivedi and Y.N. Mathur. 1985. Agro-forestry a Judicious use of desert Eco-system by Man. Division of Human Factor Studies, CAZRI, Jodhpur.

Pandey, Viash N and Pankaj S. Jain. 1991. Village woodlots in Gujarat : An experiment in Social Forestry. Working paper, IRMA, Anand.

Shankarnarayan, K.A., Harsh, L.N. and Kathju, S. 1987. Agroforestry in the arid zones of India, Agroforestry Systems 5 : 69-88 W. Jumk Publishers, Dordrechi printed in the Netherlands.

Shankarnarayan, K.A. (Ed.), 1984. Agroforestry in Arid and Semi-Arid zones. Central Arid Zone Research Institute, Jodhpur.

UNEP, 1993. Desertification control Bulletin, No. 22.

UNEP, 1991. Desertification control Bulletin, No. 19.

22

Development of Degraded Pasturelands in Arid Region — A Case Study

S.S. Rathore, B.L. Jain and B.M. Sharma

Animal husbandary is one of the main stay of farmers in arid region, but considerable decrease in the grazing lands and manifold increase in the livestock population during the last few decades have resulted in scarcity of forage. Besides, the productivity of common grazing lands has also decreased due to over grazing and impairment of ecological balance. It is therefore, essential to develop the degraded pasturelands, increase the productivity of common grazing lands and motivate the individual farmers to bring their lands under pasture to meet the fodder requirement. In view, of this, Central Arid Zone Research Institute, Jodhpur initiated a programme of pasture establishment in 1990 on degraded pasturelands (100 ha) in Tilwara, a village about 125 km from Jodhpur.

SITE CHARACTERISTICS

The village is situated at 25°55' N latitude,72° 8' E longitude. The area can be classified as arid having mean annual rainfall of 200 mm, received mainly from June to September. Maximum temperature during summer range between 40°C to 45°C and minimum during winter between 2°C to 6°C. Wind velocity remains quite high during March to June resulting in the severe wind erosion. The area is almost devoid of vegetation (having productivity of 3 to 5 q ha^{-1}) and infested with shrubs and bushes.

Soils of the project site vary from sandy loam to sandy in texture having pH 8.8, electrical conductivity 0.76 dSm^{-1} and organic carbon 0.16 per cent.

DEVELOPMENT OF SITE

No development programme can be successful without proper protection of the area from stray animals.

Trench cum mound fencing : Trench cum mound fencing has proved very economical and appreciated by the local farmers. The fencing was made by digging a trench of trapezoidal shape of 1.5 m wide at the top, 0.9 m wide at the bottom and 0.9 m deep. The excavated earth was heaped uniformly on the inner side of the trench in a trapezoidal shape having approximate width of 0.9 m at the top 1.5 m at the bottom. The edge of the mound was kept about 0.3 m away from the edge of the trench. The fencing was constructed in about 200 ha area covering an approximate total periphery of 7560 running metre length. Thus the cost of trench cum mound fencing was Rs.66,528 and Rs. 330 for one hectare.

Live fencing : For proper stabilization of mound and to check tresspass of animals, live fencing with *Acacia tortilis, Acacia senegal Prosopis cineraria* and *Prosopis juliflora* was established by direct seeding and transplanting of seedlings on trench cum mound fencing during July 1990. Direct seeding gave better results than transplanting of seedlings. Amongst the tree species established by direct seeding, the performance of *Acacia tortilis* in terms of increase in height and canopy diameter in three years was better as compared to *Acacia senegal* and *Prosopis cineraria* (Table 43).

TABLE 43. Performance of trees on trench cum mound fencing

Tree species.	Direct seeding		Transplant of seedling	
	Height (cm)	Canopy (m^2)	Height (cm)	Canopy (m^2)
Acacia tortilis.	131.0	0.97	123.0	0.93
Acacia senegal.	86.0	0.41	79.0	0.39
Prosopis cineraria.	64.0	0.13	62.0	0.13

PASTURE ESTABLISHMENT

Before establishment of pasture proper cleaning of field and soil working is essential. The soil working should be as thorough as done in case of agricultural crops i.e. the land should be properly ploughed with disc harrow and cultivator, all unwanted vegetation grubbed out and cleaned. Therefore, after completing the above operations, the programme was initiated in July, 1990 by way of establishing a high productivity grass, *Cenchrus ciliaris* strain 358, following the standard practice of pasture development. The pasture development was undertaken in phased manner. The area developed in different years and the yields obtained are given in Table 44.

TABLE 44. Area developed and yields

Year.	Area (ha)	Yield kg ha^{-1}	
		Dry forage	Seed
1990	45	2418	62
1991	35	2402	
1992	25	2420	40-60

Field studies on method of seeding and use of organic polymer for pasture establishment and fertilizer application and soil working for improvement of established pasture were conducted. Although various methods of sowing were adopted, major sowing operations were performed with cultivator as this is a simpler mode of pasture establishment under field conditions.

Sowing behind cultivator : The requisite quantity of grass seeds (@5 kg ha^{-1}) was mixed with moist sandy soil three to four times the volume of seeds, and drilled uniformly (2 to 3 cm deep) behind cultivator in 50 cm apart furrows (8 to 10 cm deep) immediately after the first effective showers.

Pelletted seed sowing : Pellets of 0.5 cm diameter were prepared by mixing seeds of grass, cow dung, clay and sand in 1:1:3:1 ratio (by volume) using sufficient quantity of water and dried. The pellets were sown in lines 50 cm apart and placed 1to 2 cm deep in the soil. Sowing operations were carried out immediately after first effective rainfall of monsoon, and also a couple of days before expected rains.

Hand sowing : Seeds were sown by hand in lines 50 cm apart in spot prepared by spade or pickaxe 2 to 3 cm deep in the soil.

Comparison of hand sowing, pellet seeding and sowing behind cultivator revealed that sowing of grass seeds behind cultivator and pellet seeding methods were superior for establishment of pasture as compared to hand sowing (Table 45). Sowing of grass seed behind cultivator was more convenient particularly for development of large area.

Use of organic polymer : The effect of seed treatment with organic polymer 'Jalshakti' on pasture establishment and forage production of *Cenchrus ciliaris* was assessed. Mixing of grass seeds with polymer 'Jalshakti' @ 5 to 20 per cent, increased the plant population, basal cover and fodder yield (Table 46).

TABLE 45. Effect of method of sowing on establishment and performance of *Cenchrus ciliaris* pasture

Method of sowing	Plant Population (000 ha^{-1})		Basal (%) Cover		Canopy cover (%)		Seed Yield (kg ha^{-1})		Dry fodder Yield (kg ha^{-1})	
	1992	1993	1992	1993	1992	1993	1992	1993	1992	1993
Behind cultivator	61.0	52.5	27.7	21.2	120.1	61.0	62.5	51.0	2010	1550
Pelleted seed	75.0	50.0	25.2	17.4	102.0	51.9	65.0	47.0	1860	1470
Hand sowing	35.0	41.2	7.5	11.3	31.3	33.9	25.0	23.0	1600	1240

TABLE 46. Effect of "Jalshakti" on productivity of *Cenchrus ciliaris*

Jalshakti (kg q^{-1} seed)	Plant Population (000 ha^{-1})		Basal cover (%)		Canopy cover (%)		Dry fodder Yield (kg ha^{-1})	
	1992	1993	1992	1993	1992	1993	1992	1993
0	22.5	41.2	4.5	18.6	7.7	32.1	1490	1260
5	38.7	48.7	15.3	15.8	71.2	52.3	2050	1510
10	47.5	58.7	22.3	23.1	92.5	84.6	2290	1670
20	63.7	61.2	26.6	30.0	126.0	87.6	2450	1760

Use of 0.25 to 0.5 kg Jalshakti with 5 kg *Cenchrus ciliaris* seed per hectare is desirable for pasture establishment.

Fertilizer use : The average productivity of *Cenchrus ciliaris* grass in arid region is low due to adverse climatic conditions and poor soil fertility. Soils are mostly deficient in nitrogen and phosphorus and therefore, application of these nutrients can play a vital role in increasing its production. Soil working once in the season can help in effective weed control and in *situ* moisture conservation. Therefore, effect of fertilizer (N and P) and soil working on the productivity of pasture established in 1990 was studied during 1992. Established pasture responded both to nitrogen and phosphorus application and response was better when combined with soil working after 25 to 30 days of growth. Application of 40 kg N + 20 kg P_2O_5 + soil working resulted in significantly higher forage yields (5400 kg ha^{-1}) and crude protein (296. 82 kg ha^{-1}) as also reflected in the other yield attributes viz. plant height, basal diameter and crown diameter (Table 47).

TABLE 47. Effect of fertilizer and interculture on growth yield and crude protein of *Cenchrus ciliaris* pasture

Treatments	Height (cm)	Basal diameter (cm)	Crown diameter (cm)	Forage yield (kg ha^{-1})	Crude protein (kg ha^{-1})
Control	98.9	19.2	40.7	2929.1	110.0
20 kg N	128.7	21.4	43.5	3283.3	143.4
20 kg N + 20 kg P_2O_5	134.6	22.6	45.9	3813.5	179.7
40 kg N + 20 kg P_2O_5	137.8	24.7	51.9	4662.5	220.5
SW*	120.2	21.5	42.0	3066.6	123.7
20 kg N + SW	148.2	22.4	45.6	3642.6	164.6
20 kg N + 20 kg P_2O_5 + SW	153.0	24.1	49.9	4350.0	209.6
40 kg N + 20 kg P_2O_5 + SW	156.2	25.4	53.8	5400.0	296.8

*Soil working

GRASS SEED COLLECTION

Seed is the most important input for any grassland development programme. Therefore, there is a great demand of *Cenchrus ciliaris* seed by development departments, agencies and farmers. Good quantity and quality of seeds were collected from pasture every year. Approximately 1500 kg grass seed of *Cenchrus ciliaris* were collected even during the first year of establishment. During 1991 and 1992, approximately 1000 kg and 450 kg seed was collected. Relatively less seed production during 1991 and 1992 was primarily due to drought. The cost of seed collection was about Rs. 5.50 kg^{-1} and seed was sold @ Rs. 35 kg^{-1}.

PERFORMANCE OF GRASS STRAINS

Performance of four strains of *Cenchrus ciliaris* (CC bundel, CC 1106, CC 75 A, CC 75 B) and *Cenchrus setigerus* (CS 76) established during 1991 each in one hectare area was compared in 1993 for their suitability in this region (Table 48).

TABLE 48. Performance of grass strains.

Strain	Height (cm)	Plant population (000 ha^{-1})	Canopy (%)	Basal cover (%)	Dry forage yield (kg ha^{-1})
CC Bundel	111.2	69.0	131.1	36.4	2020
CC 1106	111.1	68.5	120.3	34.5	1790
CC 75 A	112.8	63.5	116.2	30.5	1640
CC 75 B	114.0	62.5	116.4	31.4	1870
CS 76	104.0	55.0	85.1	20.8	1490

All the strains of *Cenchrus ciliaris* performed superior as compared to *Cenchrus setigerus* 76 and among the different strains of *C. ciliaris*, CC Bundel performed better.

PASTURE UTILIZATION

Utilization of pasture can be in two ways

(a) Harvesting, preserving and then feeding to the livestock.

(b) Grazing.

Harvesting : It is advisable not to graze animals in the pasture in first year of its establishment. Therefore, grasses were harvested and stocked in 1990 and 1992. Average dry fodder yield during 1990 and 1992 were 2418 and 2420 kg ha^{-1} respectively.

Grazing : Best way to utilize the pasture is through controlled grazing which depends upon its carring capacity. As per the standard norms 'Excellent', 'Good', 'Fair' 'Poor' and 'Very poor' condition grasslands having approximate productivity of 2000, 1500, 1000, 750 and 500 kg ha^{-1} respectively can safely provide year long grazing to 25 to 30, 20, 17, 13 and 1 to 6 adult cattle units per 100 hectare blocks respectively during the normal years. In the present case the pasture was in excellent condition which can sustain 25 to 30 adult cattle per year per 100 hactares. Accordingly grazing was allowed during 1991 and 1993 after seed collection. Approximately Rs. 2500 ha^{-1} yr^{-1} during 1991 and 1993 were realised through grazing.

ECONOMICS

The cost of development of one hectare pasture in Tilwara was as follows.

(i)	Labour charges for cleaning of site and sowing of grass seed 12 mandays @ Rs. 22 per day	264
(ii)	Tractor charges (8 hr) @ Rs. 60 per hour	480
(iii)	Inputs :	
	(a) Grass seed 6 kg @ Rs. 35 kg^{-1}	210
	(b) Jalshakti (0.5 kg ha^{-1})	50
	(c) Fertilizers (@ 40 kg N +20 kg P_2O_5)	200
	(d) Miscellaneous	40
(iv)	Fencing	330
(v)	Soil working or interculture operation once in the season.	
	Tractor charges (2 hr) @ Rs. 60 per hour	120
	Two labour charges (2 man days) @ Rs. 22 per day	44
(vi)	Maintenance and other recurring expenditure	100
(vii)	Harvesting and seed collection.	
	Harvesting charges (average production 2000 kg ha^{-1}) 10 mandays @ Rs. 22 per day	220
	Seed collection (average production 60 kg ha^{-1}) 15 mandays @ Rs. 22 per day	330
	Total (i to vii)	2388

Income :

(i)	Average fodder production 2000 kg ha^{-1} @ Rs. 1.5 kg^{-1}	3000
(ii)	Average seed production 60 kg ha^{-1} @ Rs. 35 kg^{-1}.	2100
		Total 5100

The pasture development is advantageous over crop cultivation in this region. During drought and famine years, there are complete failure of crop in this area. Therefore, establishment of pastures is the only remedy for sustained productivity and mitigating the drought effect. Further there is no recurring expenditure in pasture establishment as in crop cultivation.

RAISING OF TREES IN THE PASTURE

Introduction of tree species in the pasture lands provide not only nutritiously better fodder during the lean period but also helps in soil fertility buildup and serve as shade for animals. In addition, fodder trees and shrubs ameliorate the micro-climate and thus induce conditions for better regeneration of grasses. Many tree species are economically important for fuel and timber. With a view to develop silvipastoral system, 30 cm high ridges at 12 to 15 m distance were created by disc plough across the slope before grass seeding. Seeds of fodder and fuel tree species namely *Prosopis cineraria, Acacia senegal, Acacia tortilis, Colophospermum mopane, Harwickia binata* were sown on the ridges during 1991 and 1992 . This practice was very effective for successful establishment of trees and in preventing the runoff from field. Trees attained the height of 70 to 150 cm in two and half years.

A hortipastoral system with budded plants of ber (*Ziziphus mauritiana*) was established in 1990 in 5 hectares area but survival and growth of ber plants were poor due to lack of irrigation facilities. Further, mortality in grafted ber plants occurred due to high temperature and drought conditions.

Raising of shelterbelt : Shelterbelt is a barrier of trees intended for the conservation of soil and moisture and for the protection of forage. Shelterbelt was created by seeding two rows of tree species namely, *Acacia tortilis, A. arabica. A. senegal* and *Prosopis cineraria* in 15 ha area. Among the different tree species *Acacia arabica* attained more height (3.31 m) and canopy (6.84 m^2) as compared to *Acacia tortilis* which attained height of 2.83 m and canopy 4.36 m^2.

Tree plantation along field boundary : About 10,000 plants of *Dichrystachys nutans, Acacia tortilis* and *A. senegal* were planted along the field boundaries and roads within the project area both in grass cover and without grass (Table 49).

TABLE 49. Performance of tree species

Species	Without grass		With grass	
	Height (cm)	Canopy (m^2)	Height (cm)	Canopy (m^2)
		Sandy loam		
Dichrystachys nutans	137	1.12	—	—
Acacia tortilis	120	0.74	76	0.33
Acacia senegal	76	0.29	68	0.23
		Sandy soil		
Acacia tortilis	171	1.85	111	0.86
Acacia senegal	121	0.90	—	—

23

Transfer of Agroforestry Technology in Arid zone – A Case Study

R.N. Singh

High risk and low crop yields are major problems of arid zone farmers. As a result farmers of this region live in subsistence conditions. Agroforestry not only cover the risk but also provide food, fodder, fuel and fruit to the farmers.

SOME PROBLEMS

There are certain typical problems for successful implementation of the agroforestry programmes in the arid regions.

Animal population

Very large number of animals consisting of cow, goat, sheep and camel are not stall-fed and left for grazing in the field. Not only they damage existing vegetation, they do not allow newly planted tree seedlings or pasture to take good stands. The villagers neither individually nor collectively protect it.

Farm size

The small and the marginal farmers constitute a large proportion in the village. They have less land and so hardly afford area for growing grass.

Education

The villagers are ignorant of improved practice of agroforestry. They rarely get education from extension agent who themselves have less acquaintance with it. Largely it is due to job structure of an extension workers. It is evident from historical changes of rural development programme. Soon after independence, when community development started, the extension job was to teach villagers about crop, livestock, cooperative, health, and family planning. Their major time was spent on crop improvement. This approach continued for a long time. In early seventies, a drastic change was through introduction of the training and visit (TSV system), and extension was confined to only improving crop yield to feed increasing population. On the other hand, the forest department was exclusively for afforestation; it concentrated on forest land.

Input

Input is as important as innovation. It should be available to the villagers adequately at right time with good quality. The basic input viz. nursery for seedling are mostly in the city or nearby it. The villagers do not have interest to visit long distance located nursery. Very often, seedlings are not avaliable at right time of tree plantation and according to the villagers choice. Private nurseries often do not grow horticulture seedlings true to its variety. Grass seed is not available in the market.

Scientist

There is no consensus among the scientists about agroforestry. Some feel undesirable of growing trees with crop. The shade of trees hampers crop growth and shelters birds. The proponent consider birds - benefiting crop destroying and feeding harmful insects.

Innovation

Of late a few research institutes initiated systematic research on agroforestry. There is dearth of multidisciplinary research consisting of crop, tree and grass. Lack of understanding the efficacy of traditional agroforestry practices among the scientists lead research more as academic rather than solving the practical field condition. The exotic species (*Acacia tortilis, Prosopis juliflora*, etc.) are mostly emphasised in research because these grow fast and bring greenry in short time. The villagers have opposition to it and they like native species. These grow slow but have multifold use and compatible with crop. Farmers are so intimate with these species that ladies gave their life in saving *P. cineraria* (khejri). Finally the agroforestry innovations do not give high profit as well as quick and visible result, therefore, the villagers are not inspired with it for adoption.

Testing of innovation

There is lack of provision for large scale testing of the innovations. Testing helps in modification of innovation so as to suit local condition. Several examples can be cited from the agro-forestry extension project conducted in villages of arid Rajasthan. Plantation of nursery with improved variety of *ber* (*Ziziphus mauritiana*) could not survive as it was not possible for the farmers to irrigate even for one year due to scarcity of water. However, planting local root stock of *ber* (*Ziziphus rotunidofolia*) and budding with improved variety in succeeding year was cent per cent success. The plantation of tree seelings 10x10m spacing with crop was considered close by the farmers. They accepted plantation at two to three times of the recommended.

Apathy

The rural society is static. A great satisfaction with the present condition exists in the villages. The aspiration and urge to improve the degraded ecology is lacking. A number of benefits are derived from tree and grass which are even dwindeling, for example improved grass Dhaman (*Cenchrus ciliaris*). Individual or collective efforts to have more area under tree and grass through adoption of agroforestry are lacking.

In July 1990, 'Sar', a village 35 km from Jodhpur was adopted for extension programme on agroforestry. Both, formal and non formal methods of educating the villagers, were adopted. More emphasis was laid on the latter method of education. The following practices were introduced.

SHELTERBELT

The results obtained under shelterbelt programme are presented in Table 52. Progressive increase in the number of farmers adopting the practice of shelterbelt plantation is an indication of its popularity. The farmers are maintaining the shelterbelts well, because it protects their fields and save expenditure and hard labour on digging ditches and keeping dry thorn on the fences. Plantation of *A. tortilis* and *P. juliflora* along the fences and local species within the fields is prefered by the farmers. The well to do farmers usually go for two row shelterbelt, while the small farmers opt for one row shelterbelt. As the benefits of shelterbelt plantation have become clear almost all the villagers have now decided to adopt this practice.

TABLE 50. Shelterbelt plantation at Sar village

Year	No. of farmers benefited	No. of tree seedlings supplied	Trees survived	Pre cent survival
1990	6	1,249	688	55.08
1991	15	3,967	2110	43.18
1992	45	8.728	5966	68.35
Total	66	13,944	8764	

SILVIPASTORAL

This activity is continuing in 8 ha of community land as detailed in Table 51.

TABLE 51. Silvipastoral activity on community lands at Sar village

Spacing (m)	1991	1992
(A) 3 x 3		
Area (ha)	1	4
No. of plants	1300	3100
(B) 10 x 6		
Area (ha)	1	2
No. of plants	700	1300

NURSERY

Considering the increasing demand for tree seedlings for shelterbelt plantation, it was decided to establish a nursery in the village. Accordingly, the usefulness of nursery was explained to farmers. Five farmers were convinced of its utility and consented to follow the practice. One of them was a farm women, Mrs. Lali Devi. The Divisional Forest Officer, Jodhpur provided the free inputs (Polythene tubes and seeds) and also an incentive of Rs. 0.45 per seedling after the success of nursery.

A two day training programme was conducted on nursery management techniques. Mrs. Lali Devi contributed in the conduct of the training. In all 7000 tubes were prepared. All the operations of nursery were done swiftly and with great enthusiasm by the trainees. Motivated by the training, a farmer's leader started a nursery with two thousand polythene tube. Mrs. Lali Devi received Rs.3000 from forest department for 6500 survived plants. She bought a buffalo for her own farm. In 1993, she raised 10000 seedlings. Another villagers raised 15000 seedlings during the same year.

AGRI-HORTICULTURE

Improved varieties of *jujube* (Seb, Gola and Mundia) can give reasonable cash return from the fruits, even during the drought years. Budding of these varieties can be done on local root stock of *jujube* (*ber*) which grows naturally on the farmers fields. The budding on local root stock with improved varieties was started in 1990 (Table 52).

TABLE 52. Budding of improved varieties of *jujube* on local root stock at Sar village

Year	No.of farmers benefited	No. of root stock budded	Success of budding	Percentage of success
1990	5	160	68	42.50
1991	7	258	110	42.63
1992	22	1097	332	30.26
Total	34	1515	510	—

Several factors contributed to the low success rate of the technique, yet farmers are convinced about its utility.

GRASS CULTIVATION

Improved variety of Dhaman grass (*C. ciliaris*) CC 358 was introduced at four farmer's fields in 1992. A training was organised at the village and the importance of grass and methods of sowing were taught to 30 participants. The sowing was done three days before the onset of monsoon. It was a good rainfall year. Farmers got adequate return for their milch animals.

WATER HARVESTING

Construction of 'Tanka' a lined tank is an effective traditional water harvesting system in the Indian desert. All the 15 farmers selected under the programme were given an incentive of 650 kg cement while other expenditures were born out by the farmers themselves. The capacity of each tanka was 10,000 litres and the size was 8'2" (length) x 6'2" (width) x 5'8" (depth). The rain water stored in tankas was used for drinking, watering of plants and raising the nurseries.

COMPOST FROM FARM ORGANIC WASTE

The use of chemical fertilizers is very low in arid areas. Compost making was selected for extension work. Trainings were organised at the village and the following methodology was taught.

- Dig a pit of approximately 3 m length 1.5 to 2 m depth and 2.5 m width at a convenient place near the cattle shed.
- Spread 3 to 4 kg of farm litter per animal in the cattle shed in the evening and collect the urine soaked litter next morning.
- In the pit, spread a 30 cm layer of farm litter and urine soaked litter from cattleshed.
- Soak it with water containing 200 gm of urea per 10 litre water. Cover it with a 10 cm layer of cow dung slurry mixed soil. Spread a second 30 cm thick layer of the mixed refuse and urine soaked litter from cattle shed, soak with urea containing water, and thencover with 10 cm of cow dung slurry mixed soil. Repeat this process till the heap rises to a height of 45 cm to 60 cm above the ground level. Then soak the top with urea containing water and cover with cowdung slurry mixed soil.
- The decomposition is complete in about three to four months.